SCIENCE vs. SCRIPTURE
and
EVOLUTION vs. CREATION

SCIENCE vs. SCRIPTURE
and
EVOLUTION vs. CREATION

Brian R. Donnelly

Science vs. Scripture and Evolution vs. Creation
Copyright © 2020 by Brian R. Donnelly. All rights reserved.

No part of this publication may be reproduced, stored in a retrieval system or transmitted in any way by any means, electronic, mechanical, photocopy, recording or otherwise without the prior permission of the author except as provided by USA copyright law.

The opinions expressed by the author are not necessarily those of URLink Print and Media.

1603 Capitol Ave., Suite 310 Cheyenne, Wyoming USA 82001
1-888-980-6523 | admin@urlinkpublishing.com

URLink Print and Media is committed to excellence in the publishing industry.

Book design copyright © 2020 by URLink Print and Media. All rights reserved.

Published in the United States of America

ISBN 978-1-64753-135-5 (Paperback)
ISBN 978-1-64753-136-2 (Digital)

18.12.19

CONTENTS

Introduction 7
Chapter One: The Planet Earth
    (4.6 Billion Or 6,000 Years Old?) 9
Chapter Two: God And Creation
    (A Unique Perspective On Both) 21
Chapter Three: The Theory Of Evolution 35
Chapter Four: God And The People 52
Chapter Five: Jesus
    (Just A Man, Or Son Of God?) 66
Chapter Six: Heaven
    (What's It Like?) 80
Chapter Seven: Final Thoughts 91

INTRODUCTION

Have you wondered why there is such a huge discrepancy regarding the age of our planet? Scientists say the planet is 4.6 billion years old and the Bible says that the 'Earth' is only 6 thousand years old. Well, the scientists and the Bible are both correct about the age of the planet and of the 'Earth'. This is all explained in chapters 1 and 2 of my book which provides a totally unique and provocative analysis of creation and re-creation that combines science with Scripture in the search for truth. What is written is credibly presented with fact and unique Scripture interpretations that, combined with science, gives us a more realistic account of creation and re-creation.

In this book you will also be convinced that creation is the more viable option to believe in than evolution and the big bang theory. The book breaks down the sciences so that you can see, with clarity, what the scientists want you to believe. You will also learn more about Jesus from a perspective that is not necessarily Biblical in nature. You will get a detailed description of what Heaven is like, that is not visionary, but supported with Scripture and NDE experiences. The book also writes about how God relates to us today and how we relate to him and many more relevant and interesting topics that will satisfy your palate.

Chapter One

THE PLANET EARTH
(4.6 Billion or 6,000 Years Old?)

Scientists claim that our 'planet' is 4.6 billion years old. How they can be so precise, without actually witnessing the planets beginnings, is questionable. But for the sake of argument let us assume that the scientists are at least in the ballpark with their estimate. Creationists say that the 'Earth' is 6,000 years old. Who is right? Well, they both are. Let's look at what the Bible says first. Moses is credited with writing the first five books of the Bible or at least most of it. All Biblical scholars agree to this. I think that most believers and even non-believers agree that Moses was an incredible historical religious figure, to put it mildly, who led the Jews out of Egypt across the Red Sea (Exodus).

According to Biblical scholars Moses began writing the book of Genesis at around 1445 BC, but there is no way to be 100% sure. The Bible seems to argue for a 1445 BC date for the Exodus at which time it is believed that Moses began his writing. We know that the Lord had spoken to Moses when he gave him the words of the Covenant - The Ten Commandments. In Exodus the Bible tells us that Moses spent forty days and forty nights on Mt. Sinai. Without going further with this, keeping simplicity as our goal, it can be believed or assumed that the Lord revealed, either directly or under the Spirit's inspiration, the order of creation to Moses if not the entire book. For now, we are only concerned with the creation itself. Let us look now at what the Bible tells us about Earth's creation, Genesis, chapter 1 verses 1, 2, and 3 (King James Version).

1. In the beginning God created the Heaven and the Earth.
2. And the Earth was without form, and void, and darkness was upon the face of the deep, and the Spirit of God moved upon the face of the waters.
3. And God said let there be light: and there was light.

I read this over and over again and it became very obvious to me that the Bible is telling us God took something that was there already and renewed it, recreated it to its present state of majestic beauty and perfect diametrical form, like we would take a very old house and rebuild it anew. It dawned on me that the planet, that scientists say is 4.6 billion years old, had to be a veritable mess with all the asteroids, comets, earthquakes, tsunamis, and what have you bombarding the planet over billions of years including the Ice Ages. After what the planet had to endure over all that time it's not possible that the planet could look like it does now. Even the scientists agree that a massive asteroid hit the planet millions of years ago and along with many other factors, caused the long Ice Age or Ice Ages, killing off many of the species on the planet, including the dinosaurs. The Ice Age, according to various scientific estimates lasted a few million years until about 10,000 years ago or so, not too long before God recreated the planet and called it "Earth".

As I began to research this further I was pleased to find out that there are religious scholars who also interpret the seven days of Genesis as a record of how God regenerated the Earth after an older world order perished. Also the old Creationists biblical interpretations were of the re-creation belief. Let us now look at each verse closely. Verse 1 says: "In the beginning"—. Moses is obviously referring to the Earth that God has recreated, the Earth he is living on, the creation of animals that he knows, and the creation of modern day humans Adam and Eve and not the beginning of the original planet as many modern theologians believe. Verse 2 says: "And the Earth '**was**' without form and void."

We all know, of course, what the word 'was' means. In Hebrew the word for 'was' is hayah. It is defined as a verb that means "to exist, that is, be of, have become, come to pass." Why would God create a world without form and have nothing in it? He is describing what he started with. Verse 2 continues with, "And darkness '**was**' upon the face of the deep". To understand this, we must look again at what our scientists say about the massive asteroids and comets that hit our planet causing dirt and debris to fill the atmosphere which helped cause the Ice Age due to blocked sunlight. One such comet, called the 'Younger Dryas comet impact event', struck our planet prior to the re-creation event. We will talk about that in **chapter two**. "Without form" very simply means that throughout all of the billions of years of wear and being battered the planet lost its natural form. Go on your computers and look at pictures of our beautiful 'Earth' today from outer space and see how perfectly round (like a basketball) the planet is. Ask yourself, is this the same planet that came into being 4.6 billion years ago, untouched? Or is this the planet that God took and renewed and recreated, as the Bible tells us, to look like it does today with all its magnificent beauty all around the globe. Finally verse 2 concludes with - "And the Spirit of God moved upon the face of the waters". Again, the Bible is telling us that the water was there already. It doesn't say the Spirit of God created water upon the Earth, only that he moved upon the face of the waters.

Verse 3 says: "and God said let there be light: and there was light." Perhaps God had just cleared the atmosphere (skies) of asteroid and comet debris at that moment of time, because he wasn't talking about the creation of the Sun. This was an instantaneous event. We don't know how much Moses could really understand about the scientific nature of the planet. It's amazing that he recorded what he did without any real science available to him when he wrote Genesis. Everything he wrote is not incorrect only maybe God did not intend the Bible to be a scientific manuscript that would only confuse those who read the good book. For now we will not go any further into Genesis so that we can take a closer look at planet Earth itself. In **chapter two** we will explain in greater detail of how we arrived at certain conclusions regarding creation and the re-creation of the planet. The Bible is very clear though, that the planet was recreated and made anew by God 6,000 years ago. This is the only thing that makes sense and the only thing that can answer the planet age discrepancy dilemma. We can't dismiss the scientist's age claims. There are proven dating methods used by scientists to determine how old rocks, etc. are, maybe not precisely but accurate enough to know that the planet is very old and that is undeniable. When creationists insist that the planet is only 6,000 years old they lose credibility. Everything else regarding the planets beginnings is based on theory (The Big Bang) and can't be proven. The only written account that we have about the Earth's beginnings is recorded in the bible by the author Moses. The Bible is regarded as truth and is sacred to most believers.

How was the original planet created? We can believe, if we choose, that God created the original Universe with all the planets, Sun, Moon, stars, etc. or you can believe the scientific theory that the Universe created itself through the 'Big Bang Theory'. This is the most accepted theory by scientists. What is the Big Bang Theory? Scientists theorize that 13.8 billion years ago there was a rapid expansion of the Universe (some use the term explosion) carrying matter through space with it. We're supposed to believe that this matter then fused together over time, after a cooling period, to form the planets and Solar System billions of years later. They say the Earth itself started with a rocky core with heavy elements colliding and binding together to eventually form a perfect diametrical ball with all the rocks turning into soil and then forming all of the beautiful mountains, rivers, streams, etc. REALLY? Flying rocks fusing together to create all the planets in the Solar System while rotating around a sun that decided to be different than the other planets by remaining stationary and fiery, all by itself, while the Moon decided to show up 100 million years after Earth created itself. Unless I'm watching an episode of the 'Twilight zone' there is no way, in my right mind, that I could ever believe this 'theory' considering how complex the Earth is and how delicate the balance of life has to be for us and the planet to survive. Let's talk a little about that and how the Earth works without getting too deep into scientific data.

The Earth travels around the Sun at 67,000 miles per hour. That's about 18.6 miles per second. In 5 seconds it travels 93 miles. That's pretty fast. If you ever were in a car that was going 100 mph it's a little scary. But right now, wherever you are and whatever you're doing, you are on a planet that's hurdling through space at 67,000 mph. Not only that but the planet is also spinning at a 1,000 miles per hour and also wobbling. There are many reasons why this is happening, but to me the most important reason is that all parts of the planet get sun and all parts of the planet are habitable. The Earth makes one revolution around the Sun in a year. A slight adjustment is made every leap year. What's amazing is that it also makes one rotation (spin) in 24 hours (1 day). That means it makes 365 rotations a year which, of course, is the amount of days there are in a year. One is synchronized with the other. This is already beginning to look like an intelligent design rather than a fortuitous (occurring by chance) Big Bang happening.

Next, let us look into why we don't fly off the Earth as it spins so fast or fall off the bottom of the Earth as it speeds through space. Well, for one thing gravity keeps us grounded and being enclosed in an atmosphere, that moves along with the Earth, allows us not to feel any of the movements of the planet speeding through space. It's like being in a Jumbo Jet traveling at 500 mph. As long as the ride is going along smoothly and you're traveling at a fixed rate of speed, it feels as if you're not moving at all in the Jets compartment. When you get out of your seat to go to the bathroom you don't fly to the back of the plane. Earth is moving at a fixed rate and we and everything on the Earth is moving along with it and that's why we don't feel the Earth moving, spinning or wobbling. But, if the Earth were to slow down or speed up we would feel it. All these wonders of the Earth we take for granted because we don't think about it. We can walk out of our homes early on a beautiful sunny Sunday morning and all is perfectly still and quiet not realizing that we are barreling through space at an unbelievable rate of speed at that moment.

Another wonder of the Earth is oxygen which is needed, of course, for all living things to survive. Oxygen is everywhere on the Earth. It's amazing. You can close all of your windows and doors in your home and go into your closet and still breathe oxygen. It reaches every nook and cranny of the planet. I find that astounding. In keeping things very basic, there are (2) main sources that feed Earth its oxygen. One source is phytoplankton, one-celled plants that live at the ocean's surface and the other is green plants that live on land. When humans and animals exhale oxygen it creates carbon dioxide. All green plants, including trees and grass, need carbon dioxide along with sunlight, water, and soil to make their food. Through this process of making food plants create and release oxygen back into the air we breathe. Plants release more oxygen than they consume. An enormous amount of this oxygen is added to the air every day. Winds help blow the oxygen all around the world even at the north and south poles where there are no trees. The cycle goes on and on. We sustain the plants and the plants sustain us.

We have a clear atmosphere so that the warmth of the Sun could sustain life and plants could survive. We also have the ozone layer to protect us from ultra violet rays from the sunlight. Water is also vital in sustaining life and it is everywhere for us. Clouds form and rain falls to replenish our lakes, streams, and rivers along with melted snow in some regions. The Earth also has to be the right distance from the Sun during its journey of one revolution (365 days) around the Sun. If it wasn't we would either freeze or burn to death. Nothing goes out of kilter year after year, decade after decade, century after century, like the energizer bunny that just keeps going and going and going.

We can only be in awe of how our planet functions each day to allow us to live in a place that is also very beautiful on top of everything else. If you believe in God you should be very grateful for his gift of life and the love he has for us all. Astronauts have affectionately dubbed Earth as the "blue marble". What you cannot see from space with the naked eye is a slight bulge at the equator due to the Earth's rapid rotation at its axis. If the Earth were to stop spinning this bulge would not be present. The Earth is perfect in its creation. There is so much more we could write about the Earth's highly sophisticated workings but for our purpose enough has been said to allow us to believe that only a master planner could have designed such a complex universe.

Here is a summary of how the universe works. The Earth and the planets in our Solar System orbit around the Sun. The Moon travels around the Earth. The Sun stands still within the Solar System but the entire Solar System revolves around or orbits around the center of the Milky Way galaxy. It would take over 200 million years to make one complete orbit. We live in the Milky Way galaxy.

What do you believe? That the Earth and Universe created itself as a result of an explosion (rapid expansion) in the Universe or do you believe in an omnipotent God, the Creator, who made the Earth and Universe. Some people say 'how could God do all this' putting God on the same level as they are. 'If we can't do it, how could God do it'? In the Bible it says that "God created man in his own image", but do you think that God would have given us the same brain that he has and create billions of Gods? Some have said that we only use 10% of our brains. Technically speaking, we use all of our brain in some way, but how much of the intellectual portion of our brain do we use when comparing it to God's? God has a brain that would be unrecognizable to modern science. He also has an eternity of experience behind him to accomplish any and everything that we could ever imagine. We know that he has conquered death.

Look at what 'we' are doing now with our 'tiny' little brains in the field of regenerative medicine which deals with the "process of replacing, engineering, or regenerating human cells, tissues, or organs to restore or establish normal function". Is it a stretch to say that someday we will be able to extend life indefinitely? Only 200 years ago medicines and hospitals were just beginning to scratch the surface of effective patient care. Many who entered hospitals back then never even recovered at the hospital due to infections, diseases, etc. that they contacted at the hospital. There weren't enough medicines to take care of all the many ailments that people had. Today it's a different story. What about the next 200 years or the next 2,000 or even 10,000 years from now. What will we be able to do? It's unimaginable with technology soaring as fast as it is. Could people imagine that we would walk on the Moon 2,000 years ago. The point being here is that we shouldn't under estimate the power of God to accomplish anything he wants to.

We don't know how God does things. It's beyond our comprehension. Why can't we open our minds to the fact that God is so much smarter than we can ever hope to be. His IQ is off the charts. He makes Einstein look like an imbecile or an idiot. He probably looks down at us and says to himself "isn't that wonderful, they developed a primitive internet system." But at the same time he is probably very proud at what we have accomplished because he loves us.

How did God come into being and where did he come from? This is something that God has never revealed to us and something that we should leave alone for fear of offending him with our theories. The only thing he has said about himself is recorded in Exodus 3:14 "And God said to Moses, I AM that I AM: and he said, thus shalt thou say unto the children of Israel, I AM hath sent me unto you." I AM translated also means "I was" and "I will be." It's also wonderful that he refers to the people of Israel as 'children'. He is the supreme Father of us all.

Scientists are smart in their field of study. They are learned in their profession. But outside of their field they are just regular people like you and I and no smarter about life in general as the next person. They are not 'superior' human beings that can look into the past with certainty. They can only speculate about something that might have happened 13.8 billion years ago. There are no absolutes here. They don't happen to have any videos of the Big Bang occurrence that, according to scientists, happened 13.8 billion years ago. Think about how long ago that is. One billion years is 1,000 million years. 13.8 billion years is 13,800 million years. How could they possibly know what could have happened then? They don't! That's why it's called a theory. Even if there was a cosmic expansion of the universe, which they claim, how do they know - that created our Solar System. It's interesting how they changed the word explosion to expansion to make it more palatable for us.

The scientific community is trying so hard to deny the existence of the Creator that they had to come up with some answer as to how the universe was created, and they came up with a real head scratcher. A poll was conducted several years ago by the Pew Research for the People and the Press. According to the pole only 33% of scientists believe in God, 18% in a higher power (not God) and the rest are non-believers. Let it be noted here that not all scientists believe in the Big Bang Theory. There are those that believe in creation as well.

Scientists have an answer for everything. Who knows everything? It's common knowledge that scientists have been wrong on many occasions but their answer is always that new evidence has brought us to a different conclusion and that's how science works. As we learn more about things we have to modify our original findings. That sounds fair. But when it comes to theory (abstract thought, hypothesis, guess) we need to be very careful not to accept theory as fact and not to be impressed by scientific language that's designed to make us think theory is fact.

Here is an example of that in a scientific description of 'the Big Bang'- "If we looked at the universe one second after the Big Bang, what we would see is a 10-billion degree sea of neutrons, protons, electrons, antielectrons (positrons), photons, and neutrinos. As time went on we would see the universe cool, the neutrons either decaying into protons and electrons or combining with protons to make deuterium (an isotope) of hydrogen" and the rhetoric goes on ad nauseam. Are we supposed to accept a hypothesis so outrageous because we are impressed by scientific explanations of how something 'might' have happened or should we rely more on what makes sense, what seems more likely, or unlikely, to have happened as long as we are just hypothesizing. If a massive explosion in the Universe did occur, rocks, debris, dust, and particles would go flying out into the vast outer space, scattering and eventually dissipating into the cosmic skies. That seems more likely to have happened then for explosive matter to have collectively gathered together in outer space to form a complex solar system. Did the Solar System make itself all by chance or blind luck, so precisely that the Earth with all its complexities can function as it does to sustain life and its own existence for as long as it has? The answer is very clear that only the Creator could have designed such a masterpiece.

Brian R. Donnelly

"THE BLUE MARBLE"

Chapter Two

GOD AND CREATION
(A Unique Perspective On Both)

For something to be believable it has to make sense and be supported by fact or reasonable evidence. An example of this would be in Genesis 1:3 "And God said, let there be light: and there was light". In chapter one of this book I wrote that God cleared the atmosphere of asteroid debris. Reaching this conclusion took a lot of thought using scientific evidence along with Biblical evidence. Scientists believe that massive asteroids that have hit our planet over time helped cause the ice ages by preventing sunlight from getting through the atmosphere because of debris thrown up by impact explosions. Also during the last ice age, about 12,900 years ago, geologic evidence tells us that an enormous cosmic impact event occurred called the 'Younger Dryas comet impact' which caused global massive volcanism, tidal waves, seismic activity on a large scale, and extreme temperature changes. Upon impact, debris would have filled the atmosphere blocking sunlight and causing a state of near darkness which would eventually create an extremely cold climate. This happened not too long before the re-creation.

    Many other smaller asteroids and comets have also hit the planet. Large comets can have the same effect as asteroids.

This also from Wikipedia regarding the Younger Dryas comet impact hypothesis. "The general hypothesis states that about 12,900 BP calibrated (10,900 C uncalibrated) years ago, air bursts or impacts from a near-Earth object(s) set areas of the North American Continent on fire, disrupted climate and caused the Quaternary extinction event in North America." "The Younger Dryas ice age lasted for about 1,200 years before the climate warmed again." The fact that plant life disappeared on the planet, along with animal and hominin extinctions during the Ice Age is enough, in itself, to conclude that very little light from the Sun was evident prior to the recreation. Wikipedia goes on to describe an impact winter. "An impact winter is a hypothesized period of prolonged cold weather due to the impact of a large asteroid or comet on the Earth's surface. If an asteroid were to strike land or a shallow body of water, it would eject an enormous amount of dust, ash, and other material into the atmosphere, blocking the radiation from the sun. This would cause the global temperature to decrease drastically. If an asteroid or comet with the diameter of about 5 km (3.1 mi.) or more were to hit in a large deep body of water or explode before hitting the surface, there would still be an enormous amount of debris ejected into the atmosphere. It has been proposed that an impact winter could lead to mass extinction, wiping out many of the world's existing species."

Biblical evidence tells us in Genesis 1 verse 2 that "darkness was upon the face of the deep". Darkness does not necessarily mean total darkness, it can also be defined as partial darkness or dimness. There are other Biblical interpretations of "Darkness was upon the face of the deep" but because of its relationship to light in verse 3, darkness actually was upon the face of the 'Earth' as scientific and Biblical evidence tells us. So when Moses writes "And God said, Let there be light: and there was light" he was not referring to the creation of the Sun. Something that occurred 'instantaneously' suggests that God cleared the atmosphere (whoosh) of all accumulated atmospheric debris exposing blue skies and the bright Sun. Also Moses tells us in verse 16: "And God made two great lights; the greater light to rule the day, and the lesser light to rule the night." Here he 'is' referring to the creation of the Sun and the Moon. They could not have been created on the fourth day. The Sun was there already, prior to the recreation event. The planet is billions of years old and life could not have existed here without the Sun. Earlier, in verse two, the Bible also tells us that "The Spirit of God moved upon the face of the waters." There would be no water present in verse 2 without the Sun, only ice. So, what happened here? The logical conclusion is that God not only regenerated the planet, but he also restored and renewed a partially worn out old sun and a battered Moon. Why just re-create the planet and not the entire Solar System as well. As we continue, the writing of Genesis will unfold and become more clear for us. In verse 9 God said "Let the dry land appear," which tells us that the Earth was covered in water.

Genesis describes the planet as formless (Bombardments from asteroids, comets, and other unknown occurrences), flooded (melted ice), void (no plant or animal life) and in darkness (atmospheric debris) corresponding to our planets scientific history. It's amazing how the Bible and science intertwine. Science has unwittingly contributed to the re-creation belief.

This has been just one example of how things are being looked at in this book with a different perspective of common sense, reasoning, logic, analytics, and with whatever Biblical and scientific evidence we have. As the evidence for re-creation piles up we also have to take in account that the 7 days of Genesis is happening 6,000 years ago, not 4.6 billion years ago. It is my belief that the re-creation event occurred shortly after the last ice age. As I said in chapter 1, all of what Moses wrote did happen. Was it in great detail? Some of it, yes, but some of it wasn't. God did not go into great detail of the scientific aspects of creation for a reason. First of all, God did not intend the Bible to be written just for scholars and intellectuals but for all to read, including the wise. Also, Moses had no scientific background to process the complexity of creation science as brilliant as he was. Moses was a man, a very, very great man but only Jesus was deity. At that point in time people thought the world was flat. If Moses knew it was round they all would have known long before Columbus found it to be. How much Moses was told was limited to what God felt he could process. An example of this is found in Genesis 1:16 when Moses calls the Sun and Moon lights. We don't know if God referred to them as lights or if Moses saw them as lights. Light was very important to people at that time especially in the wilderness. Another example is found in verse 14 where Moses talks about 'lights' in the firmament of the Heaven to divide the day from the night. Here he is talking about the rising and the setting of the Sun. As the Sun is no longer visible to our eyes there is still light in the atmosphere which can sometimes be very beautiful with colors of reddish pink and blue with white and dark clouds signaling the onset of morning or evening. Moses doesn't see the Sun at those times of the day, only 'lights' across the horizon, dividing the day from the night.

It doesn't seem likely that God dictated word for word the creation to Moses. Writing at that time, from what I read, was done on clay or soft stone tablets with an iron stylus. It's more likely that God revealed the re-creation story to Moses and that he recorded what God said to him, from memory, at a later date. Genesis 1:16 ends with Moses saying "he made the stars also." This, to me, is the most important part of this verse, in that it confirms that God is the creator of the

Universe, the Sun, the original planet, the Solar System, and everything else that there is in space. While giving Moses this account it's obvious that he didn't mean that he created them on the fourth day. The 7 days of Genesis represents an enormous undertaking of God to restore his original creations. This may have been almost as difficult as the initial creation itself. The result is just absolutely astonishing. My purpose in writing the re-creation belief is to give the believer, and even those who mock the 6,000 year planet age, a cogent answer to the planet age discrepancy dilemma. When creationists say things like, 'to God, a day is a thousand years' or 'the scientists are wrong in their dating methods', people just reject or laugh at these explanations. We need to acknowledge Science and not dismiss it. By combining science with Scripture we can get closer to the truth. Getting back to Moses, we have to understand that all this information he was receiving from God must have been very hard for him to absorb and record in proper sequence. That is why he went into more detail in Genesis 2 about the creation story, leading some to believe there were two different creation accounts when there really wasn't. Genesis 1 is chronological, Genesis 2 is not. Moses added some additional information in Genesis two that he didn't record in Genesis one. This has left many to scrutinize the Bible, even suggesting that Moses did not write Genesis. There is way more proof that Moses wrote Genesis than there is proof of an unknown writer or writers contributing to Genesis. Moses did a great job recording the re-creation events considering he was under great stress at the time. What's most important is the message conveyed (the big picture) that God created the Heavens, the Earth and everything in it, the Sun, and the Universe. We can only search for the truth by analyzing all that is written and all we know, but we will never fully understand the great mysteries surrounding God's greatness. That is why Jesus told us in Mark 10:15 "To receive the kingdom of God as a little child".

I'd like to write a little about Adam and Eve and the apple. Some people mock this story as fairytale like. What they don't realize is that God had a purpose behind forbidding Adam and Eve to eat the apple. The apple was just a symbol, a test to see if the humans he just created would be obedient to him. Would humankind obey God's commands going forward for generations. Eating the apple represented serious consequences for Adam and Eve. Instead of having a life of Heaven like conditions, disobedience exposed them to not only evil but also to sorrow and all the emotions we have today. The apple in itself did not bring evil. Adam and Eve brought evil upon themselves by eating the forbidden fruit and showing that they could easily be swayed by temptation by the clever serpent (Satan). God then realized that what he had created could not be without sin. Adam and Eve was then cast from the garden. The Bible says that the wages of sin is death. God, being God, still loved what he created and eventually sent his Son to redeem our sins.

Reading interpretations about Genesis 1:6, 7, & 8 (waters above and below the firmament) I admit were somewhat confusing to me. In verse 7 some call the firmament the 'atmosphere', and that water was below the atmosphere and that water was also above the atmosphere but in verse 8 God called the firmament Heaven, not the atmosphere. I categorize this as 'beyond our understanding' and would prefer not to speculate here because there is not enough information or reasonable evidence to support a hypothesis. Instead we certainly can talk about Heaven. In the Bible Heaven is sometimes referred to as singular and sometimes as plural (Heavens). We don't need to quote Bible verses here to confirm this, but I would like to quote just one from a very reliable source. In John 14:2 Jesus says to his disciples "In my Father's house are many mansions: If it were not so, I would have told you. I go to prepare a place for you." Many mansions translated means many dwelling places. "My Father's house", of course, is Heaven. Many dwelling places could indicate that there is more than one Heaven, otherwise how would all the living who will eventually die and all who have died over the centuries fit in one place. Conventional creationist wisdom says there is only one Heaven ("my Father's house")-singular, where God himself lives, and the other heavens simply mean the atmosphere and the universe, based on Hebrew translations for a total of three Heavens. I'm not sure about that either. Is it really that important as long as we know that there is a wonderful place for us in the afterlife.

If God created Heaven at the re-creation event then where did he live before that? If the planet was really only 6,000 years old then things would be a lot simpler for us, but we know that it is what the scientists tell us that it is-billions of years old. We can't keep our heads in the sand. We have answered the planet age discrepancy very effectively with re-creation. The other mysteries we will never figure out, like why did God create this planet in the first place and not put life on it for billions of years? What if God originally created this planet for himself and his nation of people (the Angels) to live on. I thought that it was as good a guess as one could think of. I then started to expand on this a little in my mind and thought that if it was really true it would answer a lot of questions I had for myself. I began to think that after billions of years God got smarter and smarter and then created an even better place to live with perfect climate conditions, possibly one of his many dwelling places. There would be no evidence of life left on the planet during the time God lived here because nobody died. God did not destroy the planet but then experimented and amused himself with the creation of all kinds of living creatures, dinosaurs, apes, ape men, and so on until he created Adam (in his own Image) and Eve. Of course I don't know if this really happened, but at least it's as good a guess as any to answer the question of why did God create this planet in the first place. Oh the mysteries of God are abounding. Only God knows all the answers.

Why can't we see Heaven if it's above the atmosphere? It can't be very far. People who have had a 'near death experience' (NDE) or have actually died momentarily and then miraculously recovered, felt their spirit or soul leave their body and were able to see themselves on operating tables or in hospital beds and observe hospital personnel administering to them and grieving family members in waiting rooms that they possibly could not have seen in the condition they were in and where they were. There are literally millions who have had such experiences, even going back in time. In most of these cases they were drawn by a bright light or a bright light at the end of a tunnel, not with solid walls but a glowing pathway entering a glorious place seeing relatives and loved ones that have passed on before them. Some arrived in a relatively short period of time, even within the time of momentary death. Can it be that Heaven is in another dimension? Is there such a thing as another dimension? The term came from somewhere. Those who have experienced a NDE did not go to a faraway galaxy to get to Heaven.

An event that happened with Jesus can give a clue to the very possible existence of another dimension. After the resurrection Jesus suddenly appeared to his disciples in a locked upper room, out of nowhere, in the flesh. John 20:19-20 (KJV) "Then the same day at evening, being the first day of the week, when the doors were shut where the disciples were assembled, for fear of the Jews, came Jesus and stood in the midst, and saith unto them, peace be unto you". (20)

"And when he had so said, he shewed unto them his hands and his side. Then were the disciples glad, when they saw the Lord."

What's really wonderful about NDEs is that people have very vivid memories of what they experienced that have profoundly affected their lives in a positive and spiritual way. In practically all cases they didn't even want to return to their present life. These are not dreams but actual experiences. This is present day proof that there is life after death and that everything about Heaven is real. Now - here comes the scientists, the psychologists, the neuroscientists, the psychopharmacologists, telling us what these people really experienced having never experienced a NDE event themselves. This is what I mean about these guys. They can't stand hearing anything about God. Right away they have to come up with some psycho science babble to explain away NDEs, telling those who have actually experienced them that it wasn't really what they think they experienced. Really? Well, here is one scientist that actually had a near death experience herself. Joyce Hawkes, a cell biologist with a PhD, had an accident that forever changed her life and her view of science. She suffered a concussion from a falling window. The following are her remarks about her experience. "I think that part of me - that my spirit, my soul - left my body and went to another reality," she said. She was surprised at the experience. "It just was not part of the paradigm in which I lived as a scientist" Hawkes recalled. "It was a big surprise to me to have this sense of something different than the body - a consciousness different than the body - and to be in this wonderfully healing, peaceful, nurturing place." I think what I learned was that there truly is no death, that there is a change in state from a physical form to a spirit form, and that there's nothing to fear about that passage,"

She said. (ABC News). God bless Joyce Hawkes, confirming what every other person experienced during a NDE.

God's creation of humans. According to the Bible 'all' life that existed after the creation of Adam and Eve and prior to the great flood, perished in the flood except Noah, his wife, and sons, Shem, Ham, Japheth, and their wives including, also, the animals that were on the Arc. The Bible also says that waters were on the face of the 'whole Earth' which makes it clear that after the flood Noah, his wife, his sons, and their wives were the only living humans left on the planet. Noah and his family were left to repopulate the earth. God said to them as he also said to Adam and Eve "Be fruitful, and multiply, and replenish the earth." Here is where the interpretation of one word can change the whole landscape. The word earth in Hebrew could also mean land or territory - 'erets'. In English earth can be referred to as the planet or just simply 'soil' or 'ground'.

In the beginning I don't believe God intended for Adam and Eve and their descendants to populate the entire planet based on crossing oceans and getting to all corners of the planet without any means of travel except walking. Families tend to stick close to one another and not want to venture into unknown and potentially dangerous territory, like in the case of Adam and Eve's first born son Cain. We'll get to his story in a bit. My belief is that God intended Adam and Eve to populate the land and territory in which they were created where God focused his attention and where all the Bible stories are written, in that portion of the world. God took care of populating the other parts of the planet. If God populated the entire world with animals why would he just put two humans in a garden in one little place on the planet and leave the rest of the world empty of human life. The diversity of humans and cultures indicate that God put others on this planet besides Adam and Eve and the

Bible confirms this. God loves diversity in all of his creations -animals, plant life, and humans. I believe this to also be the case after the flood.

Adam and Eve had a son, their first born, Cain as the Bible tells us. The Bible is definitive when it says she again (Eve) bare his brother Abel. It is here that theologians suggest a gap, that Eve might have conceived daughters between Cain and Abel to explain how Cain could have had a wife in the land of Nod after God expelled him from the land he lived in after murdering his brother Abel. God was furious with Cain for murdering his brother. Cain was jealous of his brother because God found favor with Abel's offerings to him but not with Cain's, so Cain killed Abel. When God banned him from his land Cain became fearful and frightened to leave. God told him that he will be a fugitive and a wanderer on the earth. Cain said to the Lord in Genesis 4:13 "My punishment is greater than I can bear." Again in 4:14 he says to the Lord "It shall come to pass that everyone that findeth me shall slay me." Now it's here that the question arises - who are those that can slay Cain? Supposedly the only people on Earth were Adam, Eve, and Cain. In verse 15 God acknowledges the presence of others when he says "Therefore whosoever slayeth Cain vengeance shall be taken on him sevenfold. And the Lord set a mark upon Cain, lest any finding him should kill him." Cain then settled in the land of Nod where he took a wife.

Several reasons will be explored negating the idea of Adam and Eve having daughters after Cain and before Abel. First of all, the Bible does not say so. Secondly in Genesis 4:25 it says God granted Adam and Eve a third son named Seth in place of Abel since Cain killed him. How would that explain when God said whoever slays Cain vengeance will be taken on 'him' sevenfold. Who is 'him'? It's not another son of Adam and Eve. God certainly was not referring to Seth who came much later. Also the Bible does mention Adam and Eve having daughters and when they had them. If there were other sons and daughters born to Adam and Eve between Cain and Abel, surely it would have been mentioned in the Bible before Genesis 5:4 when it says that after the birth of Seth, Adam and Eve had other sons and daughters. It's unlikely that the Bible would neglect to mention that there were other sons between Cain and Abel even though there was a long time period between the two. If there were daughters why would they leave the safety and comfort of their parents and brother (Cain) to go off into unknown potentially dangerous territory. Remember how frightened Cain was when he was forced to leave. Also who would the daughters mate with to create those who would slay Cain. The only conclusion one can reach just based on what the Bible 'actually' says, without theological speculation, is that there were other humans on the Earth besides Adam and Eve.

Noah and his family were descendants of Adam and Eve. When they started to repopulate the land, after the flood, they were of one species. The old testament stories center around that part of the world - in what is today, Turkey, Israel, Egypt, etc., etc. We are now talking about only around 4500 years ago. Noah and family remained, for the most part, as a nation of people in that part of the world where real technology began and extended as far as Rome but never reached remote parts of the world. Why not? If descendants of Noah repopulated the entire world why didn't they also bring their technology? God focused his attention on the Israelites in that part of the world. They were his chosen people. In Deuteronomy 14:2 Moses says "For you are a holy people to the Lord your God and the Lord has chosen you to be a people for his treasured possession, out of all the peoples who are on the face of the Earth." Does it sound like the Lord wanted them to disperse and scatter throughout the entire globe and become distinctly different from one another? God loves us all equally, but I can understand at the time how he felt about the nation of Israel. On this Earth they were his first born. Did God leave the part of the world Noah lived in for Noah and his descendants to populate, and took care of the rest himself including repopulating the animals from the Arc again? After all, how were they to get to all corners of the world without Gods help like turtles, snakes, chipmunks, caterpillars, you get the idea.

I will never believe that 'one' race can turn into all the other races, cultures, and sub groups in the world, and especially in only 4500 years. I don't care how all the intellectuals try to explain that it's just skin color changes. It's not. It's facial and cultural differences also. God loves diversity and all the different types of people and animals can only come through selective creation by the Creator. Israelites look today as they looked then and mating with one another cannot turn their offspring into Chinese, African, Indian, Asian, and to every other group in the world.

However God orchestrated global population will remain a mystery. The important thing to remember is that God is responsible for the vast variety of humans and animals that are on this planet today no matter how he accomplished this enormous task. It would be boring if there was only one animal species and one human species that was gray and looked exactly like everyone else. It is amazing that no two people look exactly alike (other than twins) or have the same finger prints. How did God do that?

Evolutionary scientists have told us that life began as single cell microorganisms that eventually (over time) produced all the species and plants that exist on the planet. No matter how scientists describe their theories with fancy terms just think of what they're actually saying. Can you imagine that single cell microorganisms, so small that they can only be seen with a microscope, became the huge dinosaurs that appeared on the planet 230 million years ago, with no evidence of any developmental process' leading up to their existence. Read more about evolution in the next chapter (three).

Chapter Three

THE THEORY OF EVOLUTION

When we talk about the theory of evolution, we must always be mindful of the word 'theory'. Evolution has not proven its case and never will. This chapter will look at the theory of evolution with a perspective of logic and simplicity, and not with an endless dialogue of scientific rhetoric that will cure an insomniac. This chapter is written for anyone who wants to understand the basics of evolution. It is not written for scientific scholars, although even they may get something from this chapter. Let's start with something called 'Occam's Razor'. Occam's Razor is a principle from philosophy. Suppose there exists two explanations for an occurrence. In this case the one that requires the least speculation is usually correct. Another way of saying it is that the more assumptions you have to make, the more unlikely an explanation.

The very simplest way to look at what evolutionary scientists have told us, is to go from the beginning to present times. So, in the beginning, scientists have told us that the first forms of life were single cell microorganisms. If I had one on the tip of my index finger you wouldn't be able to see it. You would need a microscope. It's a tiny jelly type blob with no skeletal structure. Now, using Occam's Razor, let's narrow it down by rinsing out all of the scientific brainwash. What they want us to believe is that a single cell microorganism, that you cannot see with the naked eye, first became a gorilla, and then became Marilyn Monroe or Kate Upton (humans). They say this happened 'over time'. I can't tell you how many times I came across these two words during my research. This is their sales pitch, and it works. Millions have been sold on it. Anything can happen 'over time'. A walnut can become a watermelon, 'over time'.

The majority of evolutionary scientists do not believe in God. So, when asked, "If God didn't create life, then who did? Their answer is, "Life created itself." What else could they say? They then came up with a beautiful scientific word, 'abiogenesis', to make this more believable. Abiogenesis is: the original evolution of life or living organisms from inorganic or inanimate substances. Mmmm. It can also be defined as spontaneous generation. The great microbiologist Louis Pasteur gave conclusive evidence against the theory of abiogenesis or spontaneous generation. For his experiments he used a goose necked or swan necked flask. He boiled a solution of sugar and yeast for several hours. Then the flask was left unsealed for the free exchange of air with the outside environment, mimicking Earth before life. Even after several weeks there was no development of microbes in it, because the neck of the flask was shaped in such a way as to trap the outside dust particles or microbes in the neck allowing only the air to reach the solution. In this way the solution inside the flask remained germ free. By breaking the neck of the flask, Pasteur reported the development of microbes once again. The air we breathe today is loaded with living microbes. It is estimated that there's more than 1,800 types of airborne bacteria.

The experiment of Louis Pasteur clearly showed that life can only arise from a preexisting life, and that the abiogenesis theory of life is not correct. This means that present life forms can only come from preexisting life. Louis Pasteur created the law of biogenesis: Life only comes from life. This idea mirrors the Bible principles of Genesis 1: Life begets life, and like begets like. 'Like begets like' simply means that when elephants mate, they will produce another elephant, not a warthog. We'll piggyback on this later when we talk about the 'Out of Africa' theory. Evolution has a lot of theories, and we'll talk about them as well. Theorizing and hypothesizing, simply stated, means that we are guessing about something, using whatever knowledge we already have, to come to a conclusion before actually obtaining real proof or evidence. Let's quickly revisit Occam's Razor. The more you stretch out, convolute, and complicate your search for truth, the more you move away from it. The simple answer to life on Earth is that God created every living being that ever walked or crawled on this planet.

Let's talk about single cell microorganisms becoming humans or even animals. Let me be very emphatic about this. A single cell microorganism can never get to the point of developing the parts necessary to produce an ape or a human being. You CANNOT live partially developed. You cannot live with half of a head, or with a heart that has no arteries attached to it, or partially developed lungs to allow you to breathe, or an incomplete digestive system that would allow you to eat. It's IMPOSSIBLE to survive partially developed. Even us, who are complete with all our parts intact, can die if one of our organs fail. The simple proof of this is that there were never any partially developed fossils ever found in the ground, and never will be. No ½ developed or ¾ developed apes were ever found. So, how did the great apes (the hominoidea superfamily) get here 25 million years ago? They found tons of dinosaur bones who were here 230 million years ago, and became extinct 60 million years ago. Why no partially developed ape fossils found? Because there is none.

This is the Achilles heel for the case of evolution. ALL fossils of any kind that were ever found, were fully formed and complete. This could only have happened by creation and no other way.

Scientists shy away from talking about abiogenesis for good reasons, unless they have to. They concentrate mostly on apes to man. Apes to man is only a tiny fraction of the evolutionary story, considering that scientists claim that life began on Earth 3.5 billion years ago. Clearly, the apes did not just drop out of the sky. God made the dinosaurs and after they became extinct, he put the apes here. He likes creating things, that's what he does. If Gorillas (hominids) turned into humans, then gorillas should no longer exist. The scientists know this, so all they can say is that, "We share a common ancestor with apes," which tells you that they cannot identify our true ancestor, except for a single cell microorganism. Their studies involve many theories and hypotheses on how humans evolved. None of it has any legitimate proof. Later, we'll talk about how hominins (half man/half ape) appeared on this planet.

Scientists have sold the theory of evolution to millions and even caused some to abandon their religious beliefs. The evolution theory starts a long and arduous journey through time that most people do not have the patience of exploring or get bored reading about it because it's so tedious with so many dazzling words and unfamiliar scientific terminologies that impress people to conclude that these brilliant scientists must know what they're talking about. But let me point out that no human being has a crystal ball to look deep into the past and tell you unequivocally what happened millions and billions of years ago. Yes, they're smart and yes, they have found fossils that prove that other creatures have walked on this planet that weren't humans but that's the only thing that's real fact. How hominoids (apes) got here and how they evolved into humans is just pure hypothesis.

To clarify a point, I'm using the term scientist for all who are in the evolutionary field such as paleontologists, archeologists, etc. to make writing easier. Now, let's get back to fossils. There is no proof of transitional fossils, meaning that fossils cannot be connected to one another. Have you ever seen the illustration of that line of five creatures demonstrating the transition of apes to humans? The illustration shows an ape bent over walking with arms and legs on the ground and the next with arms lifted but still slightly bent and then the next walking with somewhat of a straight posture and then one walking like a human and then finally a human. For all we know all these fossil findings could have just been different species of apes, some more human-like than others, and not connected at all to one another. It's a fact that Neanderthal man and Cro-Magnon man appeared on this planet seemingly out of nowhere and then disappeared in the same manner. Here is something to think about. What if God experimented with all different types of creatures with varying degrees of intelligence before he settled on us 6,000 years ago when very highly intelligent people appeared on this planet that had never been seen before and from that point on technology has soared to such high levels that it contradicts the theory of evolution that says changes take place very slowly over millions of years.

The Cro-Magnon species (the closest to humans) appeared on the planet around 40,000 years ago in the upper Paleolithic period and became extinct during the last ice age around 10,000 years ago. They were not part of the scientist's theoretical evolutionary tree. The first hominoids appeared about 25 million years ago and hominids about 5 million. So then, it took 25 million years for a species like Cro-Magnon (who appeared on the planet out of nowhere) to develop tools that the scientists are so proud of. They refer to these tools as sophisticated when in reality they were mostly geared to hunting, fishing, cutting meats, making coverings for themselves from animal furs, and for the most part considered survival tools. They were primarily made from bone, stone, ivory, shell, antlers, wood, and were simple in design. But in just a few thousand years since the dawn of Adam and Eve we have automobiles, skyscrapers, air travel, rockets to the moon, space stations, televisions, internet system, well, you get the picture. So, which is it? Evolution travels fast or slow?

Creationists believe that God created all of life, and that all of life did not develop slowly 'over time' from single cell microorganisms. If you pin a scientist down, they will actually admit that they really don't know how life began. Charles Darwin came up with the theory of 'natural selection', stating that all species of organisms arise and develop through the natural selection of small, inherited variations that increase the individual's ability to compete, survive, and reproduce. Think about it, can single cell organisms produce the vast variety of life we have on the planet today?

Single cell organisms are the same. Remember- 'Like begets like'. Think of all the species of animals and humans that there are on this planet and then try to accept the theory that they all diversified and came into being from single celled organisms, and one became an elephant, another a duck, a mouse, a bird that flies, a lion, a hippo, a rhino, an ape, a fly, grass, a tree, and I could go on for another 50 pages. Single cell organisms along with natural selection did not produce the vast variety of life that's on our planet. Nature doesn't have a mind. Nature, of course, does not make any conscious selections or decisions needed for the survival of all the species. There is a sophisticated order of survival for humans, the animal kingdom, ocean life, insects, plants, and all the species, that could only be figured out by the Creator.

The vast variety and diversity of all the species to have the ability to survive and thrive, can only come from God.

Scientists will tell you about bacteria, archaea, eukaryotes, choanoflagellates, and hundreds of other names of cells and how they multiply but none of this proves we evolved from them. There is something called the Cambrian explosion (life's big bang) which happened about 540 million years ago. The seemingly rapid appearance of fossils in the 'primordial strata' is one of the main objections that could be made against the <u>theory of evolution</u> by natural selection. This was of great worry to Charles Darwin. The Cambrian period ended in a mass extinction. I might add that these were not animal fossils but multi-celled creatures that eventually became extinct. We must remind ourselves that if scientists say God is not the Creator they have to explain to us how everything came from nothing. They have to come up with something, no matter how bizarre it is.

Any fossils that were ever found were fully formed creatures. Coming into existence as fully formed and complete is only possible by creation. As we said earlier, most evolutionary scientists do not believe in God, so they try hard to convince us that God doesn't exist, even some creating fraudulent fossils. Scientists once reconstructed an image of a half ape and half man known as the Nebraska man from a single tooth. They later discovered that the tooth belonged to an extinct species of pig. Then there was the Piltdown man fraud. A skull was shown to be a modern human and that the jawbone and teeth were from an orangutan. The teeth had been filed down to make them look human. The bones and teeth had been chemically treated to give them the appearance of being ancient. This was of great embarrassment to the UK scientific community. Then there was the Java man, the Orce man, and Lucy. All were proven to be fraudulent. It was once believed that we were descendants of Neanderthal man. That was proven to be wrong, but what's interesting about that is there was a well developed human-like ape man that wasn't us, which leads us to believe that other fossils merely were different species of ape-like creatures that are not connected to modern man.

This notion that apes are our common ancestors just because we share common DNA with apes doesn't prove anything. 98.5 % of the genes in people and chimpanzees are identical. But a chimpanzee is a chimpanzee and a human is a human. We also share 50% DNA with bananas. Also, why isn't there anymore hominins (half man/half ape) spinning off the ape species? Right now it's just apes and humans. There's no other intermediary ape man species. To be very clear, there is no proof that any hominins ever spun off the ape species in an evolutionary lineage to humans. Another thing scientists hate to hear is the 'missing link' term. They're trying very hard to make us forget about that. Well, I'm sorry to bring that up. The 'missing link' would be the fossil of our common ancestor. It's the fossil that connects hominins to modern day humans. Scientists are eliminating the term hominid now and are just using the term hominin. They also would like to eliminate the 'missing link' term by classifying hominins as 'early humans'. I've been around long enough to remember when they used to call these creatures half man/half ape or ape men. Then they were human-like but not human. Now scientists refer to them as 'early humans'. Nice, sounds good. Helps the evolutionary tree of life theory be more believable.

Be careful with these scientists, they have to sell their theory, make it more plausible, change the names around, reclassify them, upgrade the species. Coming out of Africa 70,000 years ago to populate the world they need to be pretty far advanced. After-all 70,000 years is not a lot of time, in evolutionary terms, for genetic changes to occur in hominins from Africa (after migration) to morph into all the races and ethnic species around the planet. Remember, it's the scientists who have told us that it takes hundreds of thousands and even millions of years to make significant changes.

The next theory we will explore is the 'Out of Africa' theory. This is another one that's way out there. Even though I don't believe in the 'Out of Africa' theory, let's address it anyway. Before we get into the "Out of Africa" theory I would like to mention that there is no record of a 'missing link' fossil finding from the time of the end of the Ice Age 10,000 or less years ago, to the re-creation of the planet Earth. Notice how close the end of the Ice Age is to the re-creation. Wouldn't these fossils be much easier to find considering we can find fossils from over 200 million years ago? If we can't find the 'missing link fossils' then we have to conclude that there is no connection between hominins and humans.

Let's talk a little bit about the Ice Age or ice ages. There have been at least five documented 'major' ice ages during the planets existence. The main contributor has been the bombardment of a tremendous amount of asteroids and huge comets slamming into our planet over billions of years sending an enormous amount of debris, dust, and ash into the atmosphere, blocking the radiation of the Sun, and creating a state of near darkness, which in turn causes an impact winter, hence an eventual ice age. We get an idea of what kind of damage would occur if a huge asteroid struck Earth by looking at the asteroid that wiped out the dinosaurs some 65 million years ago. The asteroid, estimated at approx. <u>6 miles across</u> struck the planet at around 10-20 miles per second. The impact created a crater several miles deep and more than 115 miles across. The heat generated by the explosion fried and sizzled the extremely large dinosaurs to a crisp. Anything remaining alive after this event eventually died because of the ensuing extremely cold climate with no vegetation available. Obviously there hasn't been any recent sightings of any T-rexes around the planet. Next time you're driving on the highway, set your odometer for 6 miles to see how far you just traveled, and then picture, several miles ahead, an enormous asteroid, the size of the length you just traveled (6 miles), coming down at an unbelievable rate of speed. Of course, in a few seconds you would be toast. Poor dinosaurs!

Let's state some Ice Age facts. First, it was really, really cold with very little sun available, which caused a lot of snow to fall that didn't melt. Snow on top of snow eventually creates 'ice sheets'. We can't really estimate how cold it was around the world during the ice ages because the temperatures varied depending on how severe the conditions were at any particular time or place. Even scientists say that Africa was severely affected as well. Some scientists believe the hominin population may have fallen to just a few hundred, surviving only along Africa's southern coast. Here's some temperatures that have actually been recorded in present times. In 1983 Vostok, Antarctica, the temperature was minus 128.5 degrees Fahrenheit. This was the lowest temperature reliably measured on Earth. Warmer places like China, India, and Turkey have all seen temperatures at minus 50 degrees Fahrenheit. So, can you imagine the ice age temperatures?

Ice sheets on land, not the ocean, where there were enormous ice sheets. How thick or how high were ice sheets on land? Some ice sheets were up to 2 miles thick in Canada, Scandinavia, and Russia. That's 7 times higher than the Empire State Building in New York City. Seriously! Ice sheets covered most of North America. Imagine what the planet looked like after all the land glaziers melted after the last ice age. Water would be everywhere, as the Bible tells us when God said, "Let the dry land appear," prior to the re-creation. The last ice age was called the Quaternary glaciation period, also known as the Pleistocene glaciation period where ice sheets were at their maximum extent (thickness) and covered huge parts of the planet (worldwide). The Pleistocene epoch lasted approx. 2.6 million years. The true end of the ice age is not really known. Melting of the ice sheets and the devastating effects on the planet, as well as the extinctions of the megafauna, would all have to be considered. What's very interesting here is that all this brings us very close to the Bible's account of re-creation, 6,000 or 7,000 years ago. What were the Quaternary megafauna extinctions? Large animals like the Mastodons, Wooly Mammoths, Saber-tooth tigers, giant bears, etc., including the most heartiest of hominins, like Neanderthal and Cro-Magnon, all disappeared from the planet at the end of the ice age approx. 10,000 years ago. In reality, who could have survived this last ice age? It is a strong possibility, considering the megafauna extinctions, and the possible extinctions of all hominins, that the slate was wiped clean, and we started anew with the creation of modern day humans and much smaller present day animals that we know today that could only have been accomplished by God the Creator. There is no other explanation of how life could be in such great abundance in such a brief period of time. All things point to the re-creation, the planet in ruins, no plant or animal life, water everywhere, and no evidence of any intellectually developed hominins on the planet 6,000 years ago.

Scientists have said that the Bible is outdated, but in reality the Bible was way ahead of science when Moses recorded the condition of the planet over 3,400 years ago when science was just an embryo. Science is just catching up. How in the world, when you read Genesis, did Moses know what the planet looked like approx. 2,400 + years before he was born? Science supports his description of the planet to a T. There can only be one answer to the question, "How did

Moses know?" It would be someone who was an eyewitness, who actually saw the condition of the planet at the time, and that someone can only be God himself. Moses could have only received the information from Him and Him alone. This is eyewitness testimony of the Creation (re-creation) story recorded by Moses in the writing of Genesis.

The "Out of Africa" theory. Charles Darwin came up with the original theory that humans evolved in Africa where our closest living relatives, the African apes, were (Sure). Modern scientists accept Darwin's "Out of Africa" model that says all humans living today share a recent African ancestor. They have no choice but to accept this theory because the fossils of early hominids were 'all' found in Africa. They are stuck with this theory, although discoveries of early hominin remains in Israel, China, and Spain, predating the migration out of Africa, cast doubts on the theory itself. But because of reports of fake fossils, expertly crafted to appear authentic, coming out of China and dating discrepancies let's just focus on "Out of Africa". So many years of research has been spent by scientists on this theory we hate to see it go to waste.

OK, scientists, explain to us how the entire world was populated during the Ice Age by a migration out of Africa 70,000 years ago with as little as 150 to 1,000 hominins that crossed the Red Sea at the Bab-Al-Mandab Strait into the Asian continent. This was called the second dispersal, referred to as the southern route dispersal (coastal route). The first dispersal (first migration) was out of the northern part of Africa around 120,000 years ago, but they either died out or retreated back to Africa according to the scientists. Maybe they can theorize that a few got over far enough to take care of that China thing. The second dispersal then is the official migration out of Africa.

The hominins had to cross the Red Sea to get to the Asian continent. The scientists made it easier for them to do this by hypothesizing that the sea level was much lower and narrower at that time (only 8 miles wide) due to ice glaziers which tells you that Africa was pretty cold during the Ice Age. The scientists even further hypothesized that there was also a warming period then which encouraged exploration. How convenient. But how warm was it?

I even read recently that they reduced the narrowing part of the hypothesis to only 2.5 miles. This sounds much better. That would have only made it colder, not warmer. Now, let's take the middle of 8 and 2.5 and make it around 5 miles wide. This is still a long way to navigate an icy, turbulent Red Sea by what means of travel. Scientists theorize that they made rafts. Rafts that didn't get upended or leak. Rafts that broke through large sheets of ice. Remember, there is no proof of this ever even happening. But let's say they made it over to the Asian continent.

Without writing chronologically, the hominins arrived in Arabia, eventually migrated to India where some became Indian people, then onto China where their offspring became Chinese, and then onto Russia where they became Caucasians. At some point they had to make it over to Alaska from Siberia to become Eskimos (Inuits) and finally onto North and South America to become American Indians while still remaining Africans in Africa. This was some trick to pull off. How did hominins, who were one species, come out of Africa and morph into all the races and ethnic groups present today? It's not that they arrived in Arabia and mated with different species. There was no one else there to mate with. It was just them by themselves. Their offspring would be African or whatever they were classified as at the time. Remember, 'Like begets like'. Just because they moved across the water to another location doesn't mean that they then changed magically into every single group of humans in the world.

If 500 pure Chinese people moved from New York to California they would remain Chinese forever as long as they continued to mate with one another, just as an elephant will always be an elephant. This logic would apply to any group of people, Caucasians, Indian people, etc. Everything that is written here seems bizarre but it's what the scientists have told us with very lengthy scientific genetic analysis designed to make it appear plausible, but when breaking it down it exposes itself to be, not only unbelievable, but preposterous. It saddens me that young people today are accepting this teaching. It has inculcated itself into our culture like a fungus crawling into every crevice. Let's continue with the "Out of Africa" theory.

How did, whoever the heck was in Russia, get from Siberia to the North American Continent, which brings us to yet another theory. This theory is called the "Bering Strait Land Bridge" theory. Scientists liked this theory first proposed in 1590 by Jose de Acosta and widely accepted since the 1930's. Why did it take so many years to accept this theory? Because they couldn't come up with a better one. The theory suggests that during the Ice Age ocean levels dropped due to the formation of ice glaziers and exposed land that was beneath the ocean surface which conveniently became a land route from Siberia (Russia) to Alaska allowing 'whoever' to cross this bridge and populate the North American continent. This already sounds like one of those made up theories to explain how they got there. Be reminded that this wasn't the ocean floor. All the water in the ocean didn't evaporate, so there was either ocean water or ice on both sides of this land bridge. This land was above the ocean floor and then above the theoretical lower ocean levels. 60 miles of it, and not any of it under ice? They make it sound like it was nice and flat by calling it a bridge. More than likely this land, even if there was a 60 mile connection of land, had to be bumpy, rocky, icy, and not flat at all, and certainly not conducive to travel.

The Bering Strait is a strait connecting the Pacific and Arctic Oceans between Siberia and Alaska (about 60 miles long). Today it is usually frozen over from October to June. We know that Siberia and Alaska is very cold. Now, if during the Ice Age, glaziers actually formed and lowered the ocean levels it had to be a bitterly cold climate.

The migration occurred less than 20,000 years ago. Here they are running out of genetic diversity time to change into other races. Not only that but this was a very difficult time for a great northern migration. Ice sheets reached their maximum extent in this last glacial period. Accumulated snow did not melt. This created ice sheets. Travel had to be almost impossible with slipping and sliding going on with large groups of people breaking into weak points of ice into the snow. How much travel was going on at this time? How far can one get in sub zero temperatures? It's not that they had thermal boots and gloves, and clear pathways.

People die in blizzards, freeze to death with no food and cover available. Just read about people who attempted climbing Mt. Everest, even with adequate supplies and the most modern winter gear available, that have died and lost toes and fingers due to frost bite. Would children and babies and the elderly be able to endure traveling in this climate. If they made it to the middle of the ocean, winds would be whipping and swirling around with blizzards likely as well. Scientists theorize that the land bridge was 600 miles wide. That is just not believable. A 600 mile wide, fully intact, bridge with no separations? Also, how much mating was going on at over 100 degrees below zero. Scientists say they made shelters for themselves. Out of what? There were no trees there. There was only shrub tundra - dwarf shrubs such as willow and birch only an inch or so tall. They did not know how to build igloos. The first igloos ever made were in the mid 1800's by the Eskimos, or more politically correct, the Inuits. Why would you even want to travel at this time? Scientists say they were tracking animal herds. How plentiful was food? Frozen solid carcasses of dead animals buried under large sheets of ice and accumulated snow? Lots of luck trying to find food and make fire.

There is also the 'Beringian Standstill Hypothesis' which says that during the Last Glacial Maximum the migrants were trapped on the land bridge for <u>thousands of years</u> because of ice sheets and glaciers preventing them from entering North America or returning to Siberia. After conducting in-depth studies of this whole hypothesis, scientists began coming up with alternative migration routes to the south along the Pacific coast. It's doubtful that anyone on that hypothetical land bridge could have ever survived let alone reach Alaska to populate the North American continent. This is a hypothesis that even the scientists have trouble believing, that is why they are considering alternative theories.

At this point I have had enough about theories and hypothesis. Creation is far more believable then this evolutionary soup of theories. There are many witnesses for God. There are no witnesses for human evolution only that there were ape like creatures on this planet that God himself put here <u>fully formed</u>. There are no better witnesses for God than Jesus Himself, Moses, and Abram (Abraham). I know that in today's world it's not popular to quote scripture, but to make my point I need to share a few. Jesus took Peter, James, and John, his brother, to a high mountain where Jesus was transfigured and they heard God say to them "This is my beloved Son, in whom I am well pleased; hear ye Him" (Matthew 17: 1-8). You'll recall in chapter 1 of this book where God says to Moses "I AM THAT I AM: and he said, thus shalt thou say unto the children of Israel, I AM hath sent me unto you" (Exodus 3:14). In Exodus 33:11 it says "And the Lord spake unto Moses <u>face to face</u>, as a man speaketh unto his friend." In Genesis 12:2 God says to Abram "And I will make of thee a great nation, and I will bless thee, and make thy name great, and thou will be a blessing." I could quote many more passages in the Bible where God speaks to people but these should be enough to convince a reasonable person that God existed then and still exists today and will continue to exist for ever.

There is a movement going on today that is trying to discredit the Bible, saying it's mostly fictional and how do we know any of these Biblical characters even existed. How do those who bring the Bible into question know that the events recorded didn't happen. Just because the Bible is old and the writing styles are much different than how we write today does not mean that the events recorded in the bible did not happen, nor does it mean that the people in the Bible were not real people. People who disrespect the Bible have an agenda to promote. No one needs to defend the Bible. It is the greatest inspirational piece of literature that's ever been written. It is a fact based religious history book of actual events that did occur.

Back to science. I am not against science. Science is wonderful. It's made our quality of life so much better. It has extended our lives through research and medicines. Their discoveries and inventions have improved our way of life. Wouldn't we be much better served by evolutionary scientists if they put all their efforts into saving our planet from ourselves instead of going around the world endlessly digging up bones to prove the non-existence of God.

Chapter Four

GOD AND THE PEOPLE

Fast forwarding from Creation to present times, how does God relate to us today, and we to Him? Some people just can't conceive of a God with supernatural powers. They can't imagine that God can do what humanity can't, as if God is at the same level of intelligence as humans. God has created all living creatures with different levels of intelligence. Are humans the end-all, the highest intelligence in the Universe? There can't be a higher power with greater intelligence than us? Is that impossible? God has created us in his 'image' only. He did not give us His brain or His abilities to perform great feats. He has created us to be - just humans, not Gods. What God can do may not seem extraordinary to Him, but impossible to us. Jesus has demonstrated supernatural powers on Earth that were witnessed by thousands which we will cover in the next chapter. Modern humans will never see the kinds of miracles Jesus performed here on Earth. God has given us enough proof of his existence, even with the NDEs. He doesn't need to convince anybody of his existence by presenting himself to us again. If we don't believe, then that is our misfortune. Jesus told us to receive the kingdom of God as a child. He also said, "Blessed are those who have not seen and yet have believed". That's why there's such a word as 'faith'.

I had commented earlier that non-believers have dismissed the Bible as being mostly fiction. I just don't get how people think sometimes. Do people think that the Bible writers created the notion of a God just to fool the people, that all the writers had a 1500 year conspiracy to create a book that was totally fake? What would be the motivation for them? They weren't going to make a single cent for themselves from their writings. Why waste their time? Did Moses, Abraham, and all the prophets just pretend there was a God? Were all the stories about God, the prophets, Jesus, and all the disciples all lies? Wouldn't this facade, this house of cards, have come crashing down at some point? Of course this is all nonsense. The Bible is factual and trustworthy and has been a source of great comfort and spiritual inspiration for countless numbers of people.

Today God is observing the events that are occurring on Earth. He has allowed evil to exist, but have you ever wondered why evil has not been totally victorious in the world. It has been temporarily, as we see in the Middle East, but good has always prevailed over evil. That is not just a coincidence. You can go down the list of temporary evil powers in our world's history, like Hitler and the Nazis, that have been defeated or negated. God is still very much in control. As a result, though, of Adam and Eves failings we will always be subjected to evil in this world. It's not just Adam and Eve, but we would all have failed the test in the Garden of Eden. Don't expect the Earth to ever become trouble free. Earth is God's school for us. We see good, we see evil. We experience illness, depression, sadness, but also joy, happiness, and all the emotions of our human selves, good and bad. We could never really appreciate Heaven without first knowing evil, illness, depression, and sadness. It is only then, after leaving this Earth, that we can truly appreciate Heaven to its fullest, with no more pain and suffering. God knows what he is doing. He is always in charge and has a plan for us.

I have heard people say many times, "Why does God let that happen" when there is a terrible tragedy. God gets blamed for everything. God doesn't let things happen, they happen because we live in a perilous world where disasters occur. We also make bad choices (free will) that put ourselves in danger. There are many times when he does intervene, when prayers are answered, but he doesn't just sit up in Heaven and orchestrate every single event, good or bad, that occurs every single second of every single day in the entire world. There are people dying in the world at a rate of almost two per second. Do you believe that God causes every death to occur? No matter what you believe God does not make 'everything' happen in this world every moment of every day. They happen as a result of accidents, human decisions, human planning, human failures, and nature's mishaps. Is God also responsible for our unhappiness? If we're unhappy it's because 'we' are unhappy. It's not God's fault. Pray for Gods help. It will come, but not always in the way we want it to. We also have to help make it happen ourselves. As the old saying goes - 'God helps those that help themselves'. But know this - God is 'always' with us through the power of the Holy Spirit. We will talk about that in a later chapter.

One of the great tragedies in the world is when we lose a young child. I can't imagine what parents and families go through when a child is lost. It's devastating. There is one comforting thought to consider and that is that God does not look at death as we look at it. A child coming into Heaven is received with great loving care and is spared the heartaches of life on Earth and the temptations of sin and drugs. They remain pure and free of sin and are safe forever from harm and despair. They will have a greater awareness of their parents and loved ones on Earth and will look forward to a joyous reunion with them in Heaven. They will be happy playing with other children and will be in the presence of our loving Lord and other relatives that have gone on before them until the day they can embrace mom and dad and loved ones again. We only live a short time on Earth, which is just the blink of an eye to God, compared to life eternal. So parents, in your grief, stay together as your child would want you to. Don't let your loss tear you apart. Live your life and try to find happiness again. That would be the best gift that you can give to your child who is alive and well in Heaven. This applies to all of us who have lost older children, mothers, fathers, wives, husbands, and others we have loved and lost. Our separation from them is only temporary. God is not cruel but loving and kind and will not let things end for us in despair.

God is at work here on Earth. He has a purpose for each and every one of us. We would all like to be celebrities and make a lot of money, but God needs all of us to do a whole lot more important things than making movies, playing professional sports, and entertaining people. Fame does not guarantee happiness as we all know. We all would like to have more money, even the very rich. Isn't it amazing that people who make a whole lot of money never seem to have enough. How many times have we read about sports stars and celebrities who have squandered their millions. Look at corporate greed, how CEO salaries have escalated and will continue to escalate while thousands struggle to support their families. The money, the bonus's, and the perks are never enough. The love of money corrupts even the politicians who are on the take for the love of the green.

All we really need is the basics to enjoy life - a job, a roof over our heads, food, clothing, and means of travel. Keeping out of debt is 'very' important. Try to avoid running up your credit cards. Once you do that it's almost impossible to pay them back down again. Don't buy things you can't afford or that you really don't need. Your real treasures are your family and your friends. Real enjoyment comes from interaction with the people you love. Stay away from drugs and alcohol. They will bleed your pockets and make you do things that will spiral you downward. There's a real price to pay for being high or inebriated all the time. I've seen many a family destroyed because of those kinds of addictions. Don't attend concerts or go to sports venues that charge you outrageous ticket prices if you can't afford it. Your TV provides plenty of entertainment and sports to enjoy, especially with wide screen HD TVs.

I see people all the time in parks having picnics, hearing resounding laughter, with children running around and playing with one another. The simple things in life are the most enjoyable. There are so many things you can do in public that the very famous cannot do for fear of being recognized and hounded for autographs and selfies, not to mention the paparazzi. You're free to go to amusement parks, fairs, or just stroll through a quaint little town. The very famous are often prisoners in their own mansions. Yes, they have pools and game rooms but you could go swimming and play games too. The point I'm making here is to be content with who you are and with your station in life. Wishing to be famous will come with a price and not make you any happier in life. If this will matter to you at all then take comfort in the fact that God thinks you are more important to him than the very rich and famous.

If God is the CEO of Earth his first priority is to feed the masses. He needs people to work in the food industry and in the water and purification business. If you work in the food industry in any 'capacity' you are one of Gods most important people. He doesn't need you to be a movie or sports star. He needs you right where you are. Be happy that, to God, you are special. I don't care if you gather oranges from trees, or pick corn, or drive produce to super markets, or cut meats in warehouses, or fish in oceans, or work in companies that package foods, or butchers, waitresses, and fast food workers who provide meals for those that have short lunch breaks, or cashiers that move people through their registers with food carts to feed their families, all of you are vital to our survival. We can do without football, baseball, and basketball for 4 months during their off seasons but we can't live without any food on supermarket shelves for 4 months. We will then see who's really important in our society. If ever there was an organized worldwide food strike, panic of epic proportions would ensue with home invasions, pets and animals hunted down, and deaths by the thousands. Of course the likelihood of that ever happening (strike) is practically nil. But just try to picture walking into your favorite supermarket and finding absolutely nothing on the food shelves, not even an apple, and you're very hungry and every restaurant and diner is closed. A scary thought indeed.

If you work in the garbage collection business you are very important to God. Again, if no movies came out for 4 months you would survive very well, but try not having your garbage picked up for 4 months and see what happens. The smells, the insects, the wild animals, and the threat of diseases would force us out of our homes. Then there's the military, doctors, nurses, police, firemen, rehab workers, and the list goes on and on of who's really important in the daily workings of our nation. No matter how you contribute to the welfare of our society, collectively, you are all very important to God and to the success of our nation. All who are in the workforce, no matter what your profession is, and let's not forget the volunteers who do all kinds of good works, it is you who make our country the great country that it is. There is nobody greater than anyone else here on this Earth. Jesus taught us this when his disciples were arguing about who among them was the greatest (or most important) and he said to them in Mark 9:35 "Whoever wants to be first, must be last and servant of all".

I love watching sports and listening to sports talk shows. Sports are so disproportionately unimportant to the grand scheme of things and yet people live and die with the success of their local sport teams. Football has become so popular that it has supplanted baseball as our national pastime. I've heard from those who play or have played professional football say that it is a real man's game. When I watch football I don't see real men playing the game but many immature boy men out on the field. Just because you can take a hit or knock someone to the ground does not make a real man. Some men are fortunate to be born with size, strength, and athletic ability gifted by God. I would like to define, with just a few examples, what it takes to be a real man. I could fill several pages but I will highlight just some of the more important qualities. A man is a person who is not a bully. He is humble and doesn't boast. A man is kind and generous. He is someone who, if married, is faithful to his wife, active with and attentive to his children, giving up his own time to nurture them in the ways of the Lord and of the world. He gets involved (all in) in all their activities like little league baseball or girls softball. He is one who plans weekend activities with his family and not individual activities with his buddies. That's OK on occasion but family should always come first. A man always puts his family before himself. A man doesn't take what is not his. A man helps his neighbor in need. He never strikes another man unless in self-defense or in the defense of his family. A man never hits or verbally abuses his wife and children. A man lets his wife and children express themselves with attentive listening and is always affectionate with his family. John Wooden, the famous UCLA coach, once said that the best thing a father can do for his children is to love their mother. A man not only honors his mother and father but also his wife's. A man never shows weakness but displays confidence in himself and is firm in his beliefs. He doesn't waffle and is decisive, but at the same time he is willing to listen to others opinions. He doesn't whine or make excuses. A man does not take drugs or alcohol and doesn't embarrass his family by his behavior but is always in control. A man will do anything he can to keep his family together and his marriage strong so that the children never become victims of a divorce. These are just some of the attributes of a 'real' man and you don't have to be big and strong to be one.

People would like God to take care of all the problems our country faces today and make them right. God isn't about to fix our messes. He will not rescue us from ourselves if we abandon his moral teachings. We were born to have all the tools and moral codes we need to do the right things in life and yet those in power who can make things better for their fellow brothers and sisters in this country, unfortunately, only consider how they can gain more wealth and prosperity for themselves. Big money controls our nation and our political leaders as well.

The distribution of wealth in this country is alarming. As time marches on the rich will only get richer while the rest of the country will have less and less. The middle class, which thrived at one point in this country, has been shrinking for many years now. 1% of Americans have 40% of the entire nations wealth. The next 19% of Americans have 53% of the nation's wealth. Putting the two together you have the top 20% of America (the wealthiest) with 93% of the country's wealth while the 80% rest of us have only 7% of the country's wealth in which to divvy up amongst ourselves. This trend will only continue downward for us. The 80% 'us' have little power to change things to make a more equitable distribution among American citizens. We have to rely on our leaders to create a culture where big business is fair to its workforces in salaries and benefits and not be so disproportionately top heavy. They will not even attempt to tackle the problem. Our politicians will not go head to head with big business. They enable them by ignoring the dilemma.

Let's talk about corporations and CEOs. The average pay package (total compensation) of corporate CEOs is around 20 million annually. That is just the average. The CEOs make, on the average, 380 times more than the average worker. In 1978 the CEOs only made 35 times more than the average worker. The success of a company lies in its product and the people who produce that product. It doesn't take a genius to lay off thousands of workers to pump up the corporate earnings or replace pensions with 401K plans where the worker has to help fund their own pensions or send work out of the country where labor is cheaper. Jobs are increasingly being moved abroad as corporations take advantage of lower labor costs like in China and Mexico and where the product itself can be made cheaper.

The CEO should be the highest paid person in the corporation, but let me ask you this question. Are CEOs 380 times more important than, let's say, the pharmaceutical research scientists who create life saving medicines and pills that generate billions in profits for their pharmaceutical companies. These corporations were successful when their CEOs were in high school. The CEOs main function is to not screw things up in an already successful company that they had nothing to do with at the time of their hire or promotion. A pharmaceutical scientist's 'average' pay is only $150,000 as of this writing. Engineers and designers who invent and design construction equipment like excavators and cranes, etc. are only in the $100,000 range as are designers of all kinds of equipment, including toys and the brilliant Lego construction blocks. If you really think about it, who are the most important people in a company? Where would the company be without the product? Think of any product you buy - an automobile, a TV, a refrigerator, an air conditioner, and so on. All the people who invented the things we need in our lives and those that continue to improve the quality of their company's products make peanuts compared to the CEOs. All the workers in a corporation, no matter what their positions are, are valuable to the success of their companies. The distribution of compensation for all who work in corporations is so lopsided that it's disheartening.

Those who control the 'top' in the corporate world are not accountable to anyone other than themselves. There is no oversight, no unions to represent the worker. They can operate with impunity. I don't want to hear about how they have to answer to the board of directors. How's that working out so far? The superintendents of schools in this country are the equivalent of CEOs in their school systems. They are in charge of their school system. Do you know what they make? $100,000 to $300,000 a year depending on how large and where their school system is. Do you know why they only make those salaries? It's because most school systems have unions to represent their workers, to make sure that school funding is distributed fairly.

There are no unions in corporations. At one time large unions became corrupt and slowly disappeared out of the main stream workforce. Corporate workers desperately need representation today. They have a right to have voice in their workplace. If the name 'union' conjures a negative image in your mind than rename it Corporate Worker Representatives (CWR). The only workers today that get fair treatment are government workers, teachers, and those companies that have unions. They're still receiving pensions and excellent health benefits. Look at what unions did for professional athletes. The ballplayers are the workers. Some make more than the owners themselves.

Something has to be done about corporate excesses. Not only are the top pays outrageous, so are the perks. There are the corporate jets, limos, corporate boxes at sporting events, enormous expense accounts, elaborate business trips, and what have you. Worker salary increases were non-existent during the slow economy years, and yet CEO's have received an annual average of 12% pay increases during the rough economy years. Think about how much 12% is on 30 million. That's a 3 million, 600 thousand dollar raise. There should be caps on CEO salaries as there are on worker job positions. In corporate America is there any concern about how God is looking at all of this?

Corporations have reduced their workforces leaving less people to accomplish the workload. 8 hour days are now 10 to 12 hour days for exempt workers who are not covered by the fair labor standards act (FLSA). Exempt positions are excluded from minimum wage, overtime regulations and protections afforded non-exempt workers (those who generally punch a time clock). At one time it was more advantageous to be an exempt employee. You were not on the clock and received a salary generally higher than a non-exempt employee and were not docked for being late or leaving early. Today it's not so great to be an exempt employee in the corporate world. By working extra hours to complete your work, your hourly pay has been compromised when you're working 50 to 60 hour workweeks instead of 40. Companies have subtly taken advantage of this. People who work in a reduced workforce know what I'm talking about. You feel pressure from lower level managers to accomplish the department's workload with less people. They'll say something like-'we're all lucky to have a job' because the lower level managers and supervisors themselves are told by upper management to make due with less people, reorganize your workload, be creative, while they're enjoying the high life.

The board of directors of a corporation are primarily concerned with how well the stock and company earnings are performing. Do you think they're sitting around figuring out how to make things better for the average worker? There needs to be a more equitable distribution of the corporation's profits for the employees. Young workers today may not be aware that not very long ago major corporations offered their workers full company paid pensions plus a 401K plan and health benefits that were better than the HMO plans they're offering today. Eroding worker benefits have come about very slowly (little by little) which allows the employee to better adjust to less and less.

What is important is that our American citizens have a fair share of the American dream and that our children can grow up confident that there will be sufficient job opportunities for them when they finish their schooling. The government and those who hold the power in this country need to search their souls and keep jobs on American soil for the betterment of their fellow brothers and sisters. I think that's what God would want them to do. Our nation was not built for the corporate elite, but to provide equal opportunity to all working people. America is the greatest nation on Earth and deserves much better leadership from its elected officials.

Over the last few decades the U.S. Government has made a mess of our country. Our national debt is through the ceiling, the future of social security is in doubt for our children. Entitlements, Medicare, and Social Security disability fraud is so rampant it is costing taxpayers billions each year. The government doesn't know how to correct the system to prevent fraud so they just throw more money at the problem. We don't fight wars to win anymore. We are on the defensive and not on the offensive while the enemy continues to create havoc in the world. Many struggle while others thrive.

Something that can help our country's finances without hurting the middle and lower class would be to enact a flat tax for personal income earners. At this writing the highest tax bracket is 39 1/2%. Do you think that any of the mega-rich pay 39 1/2% of their personal earnings to the government? They have high priced super accountants that can create writeoffs to reduce the amount of money they are required to pay to the IRS. Even Warren Buffett, who is a decent man and mega-rich, has said that the mega-rich pay about 15% in taxes. That is 24 1/2% less than what they should pay. They absolutely 'should' pay a higher percentage to the IRS than the average worker because they can well afford to. This country has afforded them the opportunities they have to become rich. No one begrudges their wealth only their lack of gratitude for what they have by not giving enough back to society or to charitable service.

There are special tax categories for investment income that the average person is not aware of nor can he or she use because they have no money to invest, needing every cent to support their families. Warren Buffett stated that "we mega-rich continue to get our extraordinary tax breaks." When Romney ran for president, tax records showed that Romney and his wife Ann paid an effective rate of 13.9% on their 'adjusted' gross income in 2010. Kerry paid 13.1% in 2003. Before you get upset with them, keep in mind, that they probably paid more than other wealthy people did because they knew that they would have to make disclosure while running for office.

If the government reduced the personal earnings rate to 30% for the wealthy and then enacted a flat tax system that required them to actually pay that amount to the IRS, then the government would have money to start straightening out our country like paying down our national debt and insuring that Social Security is there for our children and grandchildren. The government would then be able to also reduce tax rates for the middle, lower class, and the poor so they would have more money for their families and more money to spend which would help the economy. Politicians have been talking about a flat tax for decades that would greatly simplify the tax system but no one has pulled the trigger. It doesn't take a genius to figure out why. The government will never enact a flat tax system and hurt their campaign supporters and big money backers. Many in the government are multi-rich themselves.

What we need in this country is a third party that really represents all the people including the middle and lower class. The two party system has failed us. The rich and poor are represented by our present parties but the best interests of the majority (middle and lower class) are not. It's the same-ole rhetoric with the Democrats and Republicans and nothing changes. They bicker, name call, and undermine each other which is not in the nation's best interests. Rarely does our leaders bring God to the forefront. God's values of working together and loving one another seem to be lost today. It's like they don't need His help anymore. There is nothing but negative energy coming from those who are in leadership roles, locally and at the top. I believe that a lot of our leaders actually hope that their political opponent fails, instead of trying to help him or her succeed for the sake of all the people. Let's pray for God to help us succeed in spite of the many leaders that do not have our nation's best interests at heart.

Chapter Five

JESUS
(Just a Man, or Son of God?)

Jesus. Who is Jesus? Some think he was just a man. Some think he never existed, while billions believe he is the Son of God. Those who think he was just a myth, please explain how just a myth (a name in and of itself) could have 'billions' of followers, two thousand years after the 'name' died, who pray to the 'name' every day and have a cross of a myth in their homes and in the churches all around the world. The Pew Research center reports that Christianity is the worlds' largest religion with Islam second. Why is our years defined by just a 'myth'? BC is before Christ was born and AD is after Christ was born. AD does not mean 'after death' as some people believe. AD comes from the Latin 'Anno Domini' meaning in the year of the Lord.

People who are not fortunate enough to know Jesus should not listen to the non-believers who say that Jesus never existed and that nothing in the Gospels is true. They have an agenda to deny the existence of God and Jesus by making ridiculous statements about the Bible that are unfounded. Isn't it ironic, though, that our non-believer friends have to go around with coins in their pockets that say 'IN GOD WE TRUST'. I've read where some say there is no proof, outside of the Bible, that Jesus ever existed. That is just not true. There is a reference to Jesus in the Babylonian Talmud, a collection of Jewish rabbinical writings compiled between AD 70-500. It states: On the eve of the Passover Yeshu was 'hanged'. For forty days before the execution took place, a herald—cried, "He is going forth to be stoned because he has practiced sorcery and enticed Israel to apostasy." Yeshu (or Yeshua) is how Jesus name is pronounced in Hebrew. The term 'hanged' had the same meaning as 'crucified'. Stoning was what the Jewish leaders (his arch enemies) were planning to do but became unnecessary as Pontius Pilate crucified Jesus. The Pharisees (a prominent religious Jewish sect) also accused Jesus of sorcery.

The accusations of sorcery are very significant because it points out that the Jewish religious leaders could not deny the many miracles Jesus performed, that were witnessed by thousands, so they called it sorcery (the use of magical powers that are obtained through evil spirits). Even this accusation acknowledges Jesus' great powers. We know today that sorcerers and witches do not exist but in Jesus' time people believed in such things. In one instance Jesus healed a man born blind that everybody, including the Jewish leaders themselves, knew of this man's blindness at birth. Even the fact that it was written that Jesus was enticing Israel to apostasy (the abandonment or renunciation of a religious or political belief) is a testament to how powerful his teaching was. They were afraid that Jesus could actually bring down such a powerful religion as Judaism, firmly established long before Jesus was born. Of course Jesus had no intention of doing that. He came to save us from our sins and to teach us what's really important - the love and compassion of our fellow human beings and to bring change to a religion that he thought needed change.

In Rome, in the year 93, Josephus (a well known historian) published his lengthy history of the Jews. Josephus himself was a devout religious Jewish man. His account of Jesus is called the 'Testimonium Flavianum'. He writes the following: "About this time there lived Jesus, a wise man, if indeed one ought to call him a man. For he is one who performed surprising deeds and was a teacher of such people as accept the truth gladly. He won over many Jews and many of the Greeks. He was the Messiah. And when upon the accusation of the principal men among us, Pilate condemned him to a cross, those who had first come to love him did not cease. He appeared to them spending a third day restored to life, for the Prophets of God had foretold these things and a thousand other marvels about him. And the tribe of the Christians, so called after him, has still to this day not disappeared."

This account of Jesus by Josephus came under great scrutiny by the Jewish community. They said that a Christian probably altered Josephus' writing of Jesus because they didn't believe that the paragraph could have been written by a Jewish man. Others say the paragraph is essentially authentic and supports the objective historical writings of Josephus. Josephus was open minded and was not afraid to write things that offended Jews, which he often did with his writings. Perhaps Josephus was deeply moved by Jesus and his teachings. There is no proof that the paragraph was ever altered by another writer. If a Christian writer was to alter Josephus' account of Jesus wouldn't he make it a little more subtle so as not to appear doctored? Wouldn't that undermine his goal? The point here is, whether it was doctored or not, the controversy itself is proof enough that Jesus 'was' mentioned in Josephus' writings and that is all we need to know for our purpose. Thirdly, the Bible itself is a religious 'history book' written by very honest and credible writers.

The only conclusion one can reach is that Jesus does exist and is not 'just a man' who once existed, but 'is' the Son of God.

For those who are not familiar with Jesus, a brief history is in order. If you want to know more then please read the four Gospels in the New Testament of the Bible. Jesus came to mankind by what is known as the 'Immaculate Conception', the birth of baby Jesus (the Son of God) by the Blessed Virgin Mary. Why was he sent here by God the Father? He came to atone for our sins. The Bible tells us that 'the wages of sin is death'. God in his infinite love for us could not, by his law, put his beloved children into eternal death. So God sent his only begotten Son Jesus Christ to give up 'his life', here on Earth, for all of us for the remission (forgiveness) of all our sins. He paid the ultimate price (the penalty) for all our sins, dying in place of us, in an excruciatingly slow death on the cross in a manner befitting all our earthly sins so that we may gain eternal life.

All God asks in return is for us to believe this and be grateful. Is that too much to ask of us, to believe in Jesus so that we will not perish but have everlasting life as the Bible tells us? We are saved by grace alone through Jesus' suffering and death on the cross as an earthly 'human being' feeling all the pain and mental anguish that any of us would have felt. We cannot do anything to earn our way into Heaven. Jesus did it all. That doesn't mean we can sin freely. We must be repentant and try to live a life pleasing to God and not disrespect the great sacrifice of his Son, our Lord Jesus Christ.

We don't need to be just grateful but we need to worship the Father, the Son, and the Holy Spirit. God says, in His first two commandments, that "You shall have no other Gods before me". "You shall not bow down to them nor serve them". "For I, the Lord your God, am a jealous God". Other Gods does not necessarily mean a 'being' or statues, or symbols of Gods, but can mean anything you believe in that gave you life other than God himself. An example of this is believing in evolution, the serving of a theory that says God is not our Creator or the creator of Earth, that something other than God is responsible for who we are and everything we have. This, therefore, is the breaking of the first two commandments.

Jesus not only came to atone for our sins but he also came to teach, to be an example of how we should live and love one another. And of course, after his death he rose again to be seen by many so that we may also have the hope of eternal life by believing in him.

The first four books of the New Testament are called the Gospels. There are no real contradictions among the four Gospels. They complement each other and provide a complete picture of Jesus' life and ministry. Of course, when four different authors write about the same person, each account will highlight different things about the subject that seem special, or important, to them and leave out things that the others have included in their writings. Some have tried to make an issue of this when there really is none. The four Gospels are written by Matthew (the Apostle), Mark (the companion of Peter), Luke (the companion of Paul), and John (the Apostle).

Jesus as a child grew up in Nazareth. He became strong in spirit and was a very wise and intelligent child. He lived with his mother (Mary) and stepfather (Joseph) and several younger siblings. At age 12 he was already discussing Scripture with scholars at the temple. Not much is known about Jesus growing into manhood other than helping his stepfather with carpentry and stone masonry. There is no doubt that he was probably very good at those trades also. So let's fast forward to the well known account of his baptism by John the Baptist. After his baptism and his 40 day fast and temptation by Satan, Jesus came to Galilee where he began his ministry at the age of 30. How long was his ministry? There is no definitive time frame of Jesus' ministry. Some scholars claim only 1 year, some claim 3 1/2 years. Without going into the arguments for both, others have made a good argument for more than 2, but less than 3. So let's make an approx. guess of about 2 1/2 years. Galilee is the area north of Jerusalem. Jesus went all around Galilee teaching in the Synagogues, preaching the Gospel of the Kingdom of God, and healing all kinds of sicknesses and diseases among the people everywhere he went and performed many miracles. Without listing all of Jesus' miracles the following are among the most well known:

* Born to a virgin.
* Made wine from water at a Cana wedding.
* Healed a nobleman's son in Cana.
* In front of a great crowd Jesus made many fish appear in the nets of fishermen who were out all day without catching anything. This miracle made the fishermen - Simon Peter, Andrew, James, and John leave their boats and nets and follow Jesus after Jesus said "from now on you will fish for people."
* Cleanses a man with leprosy.
* Heals a paralytic.
* Heals a man's withered hand.
* Raised a widow's son in Nain.
* Calmed a fierce storm with just a command.
* Raised Jairus' daughter to life.
* Healed two blind men.
* Healed a man unable to speak.
* Fed 5,000-10,000 people who came to see him with just five loaves of barley bread and two small fish. People were astonished at the fact that bread

and fish kept appearing in baskets out of nowhere. They were chilled with awe at the lord Jesus.
* Walks on water.
* Heals many sick in Gennesaret.
* Heals a man that was born blind.
* Cleanses ten lepers.
* Raises Lazarus from the dead
* Heals a servants severed ear.
* Second miraculous catch of fish.
* Jesus' Resurrection (this and his birth are the two greatest miracles). If you want to know the detailed stories of these miracles you can pull each one up on your computer.

Other amazing attributes of Jesus is the fact that he was able to know what you were thinking and knew what was going to happen before it actually happened. If you want to think that Jesus made these things happen then that is just as amazing.

Jesus was destined to die. It was foretold by the Prophets and commissioned by God. Jesus was arrested and charged with blasphemy by the Sanhedrin and its president and high priest Joseph Caiaphas. Caiaphas was in charge of the temple treasury, and controlled the temple police and lower ranking file. He did not want Jesus to threaten any of his power. He wanted him out of the way. He convinced the council to vote in favor of having Jesus put to death.

The Sanhedrin was the Jewish court, in ancient Israel. They could impose the death penalty, but during New Testament times, could not execute Jesus themselves. Only the Romans had that power. In summary, they finally convinced the Roman Governor, Pontius Pilate, to crucify Jesus. The crucifixion did not take place in Rome but in Golgotha, a place outside the city of Jerusalem.

Jesus said of his life "No one takes it from me, but I lay it down of my own accord. I have the authority to lay it down and the authority to take it up again. This command I received from my Father."

So it happened - Jesus was crucified by the order of Pontius Pilate on, what we celebrate as, Good Friday. On the third day (Easter Sunday) he rose again from the dead. He was nowhere to be found. The tomb was empty despite the fact that it was guarded by Roman centurions and a huge boulder blocking the entrance. The boulder was removed and Jesus was no longer in the tomb but had risen and emerged alive and well.

It's worth repeating again in this chapter, in greater detail, the account of Jesus appearing to the disciples the day after he rose from the dead. "Then, the same day at evening, being the first day of the week, when the doors were shut where the disciples were assembled, for fear of the Jews, Jesus came and stood in the midst, and said to them "Peace be with you". When he had said this, He showed them His hands and His side. Then the disciples were glad when they saw the Lord. So Jesus said to them again "Peace to you! As the Father has sent Me, I also send you". And when He said this He breathed on them and said to them "Receive the Holy Spirit. If you forgive the sins of any, they are forgiven them; if you retain the sins of any, they are retained." Now Thomas, called the Twin, one of the twelve, was not with them when Jesus came. The other disciples therefore said to him, "We have seen the Lord." So he said to them, "Unless I see in His hands the print of the nails, and put my finger into the print of the nails, and put my hand into His side, I will not believe." And after eight days His disciples were again inside, and Thomas with them. Jesus came, the doors being shut, and stood in the midst, and said, "Peace to you!". Then He said to Thomas, "Reach your finger here, and look at My hands; and reach your hand here, and put it into My side. Do not be unbelieving, but believing". And Thomas answered and said to Him, "my Lord and my God!" Jesus said to him "Thomas, because you have seen Me, you have believed, Blessed are those who have not seen and yet have believed." John 20:19-29 (NKJV)

The term 'doubting Thomas' has been used throughout the centuries whenever someone doesn't believe what another says. What's also very telling about the doubting Thomas story is Jesus' amazing telepathic powers. Without being present when Thomas doubted that the disciples saw the Lord, Jesus knew, without being told, exactly what Thomas said to the disciples when he reappeared to them eight days later.

The names of the twelve disciples/apostles are: Simon (Peter), Andrew, John, Phillip, Bartholomew, Thomas, Matthew, James (son of Zebedee), James (son of Alphaeus), Thaddaeus, Simon (the zealot), and Judas Iscariot (the betrayer) who was replaced by Matthias. Although not part of the twelve, Paul was a faithful servant of the Lord Jesus and did as much as any to further the Kingdom of God.

Before Jesus ascended into Heaven, He gave the disciples the 'Great Commission' - "Go therefore and make disciples of all nations, baptizing them in the name of the Father and of the Son and of the Holy Spirit, teaching them to observe all things that I have commanded you; and lo I am with you always, even to the very end of the age." – Matthew 28: 16-20 (NKJV)

The twelve disciples, who were just ordinary working men, spread the Gospel, with the help of the Holy Spirit, far and wide. Again here is another example of how Jesus trusted the ordinary and the humble to spread the Gospel, and what a Job they did. They suffered greatly for their faith and in most cases met violent deaths on account of their bold witness. To relate just two such deaths - Peter was crucified upside down at his request, because he didn't feel worthy to die the same way as did Jesus, and Paul was beheaded. Paul spent many years in prison writing his good works, inspiring many to become believers. They were both martyred in Rome around 66 AD during the persecution under Emperor Nero. John is the only one of the twelve generally thought to have died of old age.

How could all of this happen if Jesus truly wasn't the Son of God and the Holy Spirit was not involved in the growth of Christianity. How could a 30 year old, who walked around a small area of the planet without any means of travel to facilitate his ministry and also with no TV, radio, or newspapers to carry his message to all the people, for only a 'couple of years', capture the hearts and minds of billions who worship and pray to him every day, 'two thousand years later' if he was not the Son of God? Who listens to a 30 year old? If you're a 30 year old try walking around New York City telling everyone that you're the second Son of God. See how many followers you can gather, that their descendants will pray to you into the year 4,000. How did Jesus get the people's attention? It started with miracles, real miracles, not street magic or illusions. People came to see who this man was and then when he began to preach to them, he spoke with such authority and wisdom that they were amazed at his teaching. He was something to behold. He was greater than special. How blessed were those who actually saw him in the flesh. There was no human like him. He captivated all who heard his words.

People during Jesus time were very smart as far as God and religion were concerned. They didn't have the distractions that we have today. Their lives centered around their faith and beliefs. If you're in doubt as to the intelligence of people during Jesus' time just read the Bible and see the wisdom that's present in the Holy Book. Read the wise sayings of King Solomon in the book of Proverbs. People were not gullible as nonbelievers would like us to think and neither are the billions of people living and dead who have believed in the Lord Jesus. It's those who do not believe that have missed the truth and are the ones who need to re-examine their beliefs about God and Jesus for their own sake and salvation.

Jesus' ministry grew in spite of fierce opposition. The disciples were in disarray after Jesus' crucifixion and ready to return to their lives and jobs until Jesus appeared to them after his resurrection. Seeing him again emboldened them to go out as faithful witnesses, under great danger, to bring the Gospel of Jesus to many over their lifetimes. Matthew, Mark, Luke, John, Peter, Paul, James, and Jude also wrote so beautifully in the Bible without any real experience as writers. If you're inclined to think that they had the help of the Holy Spirit then that's a good thought. Who is the Holy Spirit? The Holy Spirit is the third person of the Trinity. He has great powers and assists God and Jesus in carrying out the mission of instilling faith and truth within the true believers by residing in us. He helps us to decide against sin and to follow the path of God and Jesus. But we still make decisions based on our own 'free will' choices. That is why, although having the Holy Spirit within us, we still commit sin. The Holy Spirit is present in our daily lives in the form of a spirit, not flesh. He is the breath of God and Jesus in our hearts and minds. He also spreads the Gospel to those who may want to accept Jesus Christ as their own personal savior. He is everywhere. How is that possible? There we go, thinking like humans again.

Jesus told the disciples that he will return again with power and great glory. He said, though, that the day and hour no one knows but my Father only. It will be the greatest event the world has ever seen. God sent his Son to us once and Jesus says God will send him again. You can count on it. Those Theologians who have tried to predict the second coming of Jesus have made themselves to look foolish. Some predict that we are in the end times 'now'. Maybe so, but how long will it be before God pulls the trigger. It could be later rather than sooner. The key to the answer lies in how sinful mankind becomes. The Bible says that before the second coming of Christ there will be an increase in wickedness, along with other things that will occur. We have no better examples of God's wrath against sin and wickedness than the great flood of Noah's time where God obliterated the Earth of all life and the destruction of the sinful cities of Sodom and Gomorrah.

If we read 2 Timothy 3:1-5 (NIV) we may very well be in the end times as some say. Let's look at what the Bible says: "But mark this: There will be terrible times in the last days. People will be lovers of themselves, lovers of money, boastful, proud, abusive, disobedient to their parents, ungrateful, unholy, without love, unforgiving, slanderous, without self-control, brutal, not lovers of the good, treacherous, rash, conceited, lovers of pleasure rather than lovers of God - having a form of godliness, but denying its power." Sounds like we're not that far away from Timothy's prophecy.

Of course no one can fathom a guess as to when Jesus will return. What will happen when he does return? There is so much about this in the Bible that it becomes very complicated. For example just try reading 'Revelation' in the Bible and you'll get an idea of what I'm talking about. So, I'm going to try to make it as simple as I can without going deep into Bible verse after Bible verse. Some believe that Jesus will reign for a thousand years here on Earth as a king which will then bring about the end of this age and a new era will begin. Then, after a final judgment by God, the end of the world will occur. Without going into the reasons why, many Scholars do not accept this interpretation of the Scriptures.

Many believe, as I do, that when Jesus returns there will not be a thousand year reign. The explanation of this involves an extensive study of all Scripture verses regarding Jesus' return. That, we are not going to venture into. Instead I am going to give a very brief summary of what many believe:

- The dead will be raised. - This requires an explanation. When you die (if you're a true believer) your soul takes leave of your body and ascends to Heaven to be in the presence of the Lord (absent from the body and at home with the Lord). When Jesus returns to Earth, at the resurrection, He brings our spirits with Him, resurrects our bodies, and reunites our spirits with our bodies which will be new, glorious, and immortal. Those who are alive at the time will be caught up together with them in the air and will be forever changed, as well, with new incorruptible bodies. We will all then return to Heaven, whole and complete, to forever be with the Lord. (Read: 1 Thessalonians 4:16-17, and 1 Corinthians 15: 51-52). The condemned, when they die, their spirits descend into Hades (Hell) to await final judgment.
- Christ's Kingdom will be released to God. Satan and all the forces of evil are destroyed.
- The world will be judged.
- The world will then be destroyed by fire.

Christ's return along with the rapture (a term many use for the raising of the dead) and the final judgment, etc. can make a book of its own. My purpose is to inform the reader in a basic understandable context the wonderful story of our

Lord Jesus Christ. Many people, including church goers, have very limited Bible knowledge, but like Jesus said, have a wonderful childlike faith with a belief system that is very solid. I would encourage anyone who is inspired to learn more about Jesus to please read the four Gospels and if the Gospels grab your attention go further into the new testament, it is wonderful reading. If you're on the fence, jump to the side of Jesus. He will welcome you with open arms.

I'd like to conclude this chapter with just one more thought regarding the credibility of the twelve disciples. There couldn't be any better witnesses for Jesus' resurrection than this group of holy men. Their honest witness of Jesus Christ is beyond reproach.

When a group of people claim that an extraordinary event has happened involving themselves, that 'wasn't true', at some point, one or more from the group, before they die, 'will' confess and say that the event never happened. But not one of the twelve disciples ever denied their witness of Jesus Christ, even dying for the cause of Christianity.

Chapter Six

HEAVEN
(What's It Like?)

What is Heaven like? A brief summary of the biblical account of the Garden of Eden (before Adam and Eve ate the apple) gives us a clue of what Heaven is like. The Lord God had planted a garden eastward in Eden. He made trees grow that were beautiful to look at and also good for food. A river flowed out of Eden to water the garden. There it divided into four rivers. God put Adam in the garden to work and keep it and to live there with Eve. This garden was not like your garden in your back yard, it was over a large piece of land. God told Adam and Eve that they could eat the fruit of any tree except the tree of the knowledge of good and evil. Initially, Adam and Eve were without clothes and were 'not' ashamed. The Lord God brought all the animals he created to Adam so he could name them. The animals were tame and not ferocious. So much is said here about Heaven in the account of the Garden of Eden.

God had intended for Adam and Eve to live in Heaven-like conditions. The Garden of Eden contained all that is beautiful to their eyes. Trees, water, grass, plants, and undoubtedly flowers of beautiful colors. Eden was God's paradise, more beautiful than any other place on Earth. God intended for Adam and Eve to eat the fruit of all the trees in the garden, except one. It was not intended for Adam and Eve to eat meat by killing the animals in the garden. The animals were tame and Adam and Eve did not fear them. Lions laid side by side with lambs and peace permeated throughout the garden. So will it be in Heaven. God did not intend for Adam and Eve to be idle by giving Adam the task of caring for the garden. So will it be in Heaven where Jesus will give us wonderful tasks to perform that would be unlike our work here on Earth with tension, stress, and deadlines. Our work will be joyous and rewarding in Heaven by serving and pleasing our Lord. We don't know what those tasks will be but it will involve service to our Lord in some way. More on that later. The fact that Adam and Eve were unclothed and did not feel any shame indicates to us that the unpleasant feelings and emotions we carried on Earth will no longer be present in Heaven. That is supported by revelation 21: 4 "And God shall wipe away all tears from their eyes; and there shall be no more death, neither sorrow, nor crying, neither shall there be any more pain: for the former things are passed away."

People who have had NDE's and had a glimpse of Heaven cannot even begin to describe the beauty and colors of this glorious place. One even said "The aroma and sweet smell of Heaven is unlike anything I've ever experienced." We know what Earth is like, the parts that haven't been touched by humankind. There are parts of the Earth that are stunningly beautiful. Mountains, valleys, ranges, foliage, flowers, beaches and oceans are all wonderful to look at. Picture Heaven as infinitely more beautiful than the most beautiful parts of our planet with colors that we didn't even know existed.

I've heard people say, 'won't we get bored living forever in Heaven'? The answer is, we will not know or feel boredom.

It will not be in our DNA. God made us to feel boredom on Earth, so we would not be idle, but hopefully productive. We will not be complacent or bored in Heaven. Ever! Here are just some of the negative feelings and emotions we will shed in Heaven. -*sorrow * boredom *hate *jealousy *sadness *depression *anxiety *tension *hurt feelings *lying *fear *worry *doubt *blame *crying *deceit *conceit *narcissism *arrogance *unhappiness *boastful *shame *lust *complacency *insecure *shy *demanding *haughty *mourning *critical *insincere *anger *self-conscious *guilt *disgust *naivety *sarcasm *superiority *self-indulgence *temptation *vengeance/revenge *untrustworthy *heartache *inadequacy *remorseful * the desire to be bad or to do evil *steal *kill *the desire to harm another (mentally or physically) *feelings of worthlessness *feelings of being un-liked or unloved *sociopathic *any kind of psychological disabilities (bi-polar, schizophrenia, etc.). There's more, but I'm sure you get the point. In Heaven there will be no more negative feelings or emotions, only the good ones. I'm sure you don't want me to list those also. You can think them up yourselves. There will be no more earthquakes, floods, volcanic eruptions, cyclones, hurricanes, tornadoes, or any other natural disasters in Heaven. We will never be hot or cold. Temperatures will be perfect.

What will our bodies be like? Our bodies will be indestructible. They will be similar to our earthly bodies but will have some differences. Our bodies will be beautiful and perfect. We will keep our identities. Man will be man and woman will be woman. Our bodies will not ache or feel pain. We will always feel good. No headaches, stomachaches, nausea, dizziness, fatigue, etc. We will never get sick. Those who couldn't walk on Earth will be made whole and will be able to do all the things that others can do. Those who lost limbs will have all the parts that the others will have. If any freak accident should happen to our body's we will instantly be repaired. The blind will see. The deaf will hear. The mute shall talk. The bald will have hair. I'm looking forward to that. No one will be shy or self-conscious or insecure. No negative emotions at all, like we said before. We will be happy and joyful. All will like to be with others and enjoy their company. No one will be disliked, but all shall be loved.

It's likely that we will be clothed in Heaven. Angelic beings are described in the Bible as wearing some kind of garments. The Angel guarding Jesus' tomb is described as wearing garments. "His appearance was like lightning, and his clothes were white as snow." (Matthew 28:3 NIV). In Revelation 3:5 (NIV) it says "The one who is victorious will, like them, be dressed in white"- meaning - he who is risen in Christ will be dressed in white. There is enough said in the Bible to come to the conclusion that we will be clothed, rather than unclothed, in Heaven.

Will we eat and drink in Heaven? Jesus said, when celebrating the Passover with the disciples: "Verily I say unto you, I will drink no more of the fruit of the vine until that day that I drink it new in the Kingdom of God." (Mark 14:25 KJV). In the Garden of Eden, God said to Adam and Eve (Before they ate the apple), "Behold, I have given you every herb bearing seed, which is upon the face of all the earth, and every tree, in which is the fruit of a tree yielding seed, to you it shall be for meat." (Genesis 1: 29 KJV). In Genesis 1: 30 (KJV) God tells us what the animals would eat: "And to every beast of the earth, and to every fowl of the air, and to everything that creepeth upon the earth, where there is life, I have given every green herb for 'meat': and it was so." The conclusion we can come to is yes, we will eat and drink in Heaven.

In Heaven, food will no longer be essential to sustain life. When we eat it will be for enjoyment and in fellowship with one another. Our new bodies will process and dissolve foods within the body, in such a way, that we would not need to expel waste. Animal meat will not be eaten in Heaven. We will not kill or sacrifice animals. Lions and tigers and other ferocious animals will be tame and not feared. Won't it be great to pet lions and tigers.

Let's take a step back and explore the possibility of injuries occurring in Heaven. I said earlier that if a freak accident, one happening out of the ordinary, were to occur, our bodies would be instantly repaired. We need to say more about this subject. I stopped short of saying we will 'never' get injured in Heaven only because of how long we will live, trillions and trillions of years and beyond (for eternity). The Apostle Paul suggests, in not so many words, that our bodies will not become injured. I dare not go against what Paul said, but only to address the 'possibility' of a freak (out of the ordinary) injury, which may or may not ever occur in Heaven. An example would be someone accidently poking someone severely in the eye. I believe that the eye would immediately repair itself.

I'm not talking about ankle sprains, knee injuries, muscle pulls, fractures, etc. that athletes commonly get. We will never occur those types of injuries in Heaven. Our bodies will be like that of Superman, who, if hit by a speeding train, the train would be demolished and Superman would just walk away unharmed. Now, I'm just being facetious, but we will be indestructible as the Apostle Paul says. Also we will have a greater awareness we did not possess on Earth to avoid injuries.

To be supremely confident of our immortality in Heaven, we need only to look at the story of Lazarus, who Jesus raised from the dead. Even with Lazarus' weak and destructible earthly body Jesus was able to repair, in totality, his dead body. Lazarus was dead for four days before Jesus arrived at his tomb. A person dead for four days would have a body completely destroyed. Cells would die because of a lack of oxygen. Brain cells would die in a matter of minutes. Decomposing starts almost immediately. Skin changes colors. Body extremities will naturally turn blue. Organs will have all shut down and rigor mortis would occur. Needless to say Lazarus was a rotting mess. Now, what happens next gives me the chills.

Jesus commands - "Lazarus come out." In a matter of seconds Lazarus emerges from his tomb completely whole and in total repair. This is really an amazing miracle Jesus performed that was witnessed by many. I wanted to review this miracle with a little greater depth so that we can all realize the enormous power the Lord possesses and what he is capable of doing. Is there any doubt then, that our Heavenly bodies will be imperishable and indestructible as the Apostle Paul has written. Jesus also raised Jairus' daughter and a widow's son from the dead.

Let's continue with more about Heaven. There will be no more procreation in Heaven. Since nobody dies, there will be no need to repopulate. Will there be relations between man and woman? Of course, but it will not be like it was on Earth. Everything will be done out of pure love and not carnal lust. Relationships will have a spiritual connection attached to it and will be more satisfying in ways we never experienced before. Husbands will reunite with wives, and wives with husbands, etc. in a different relationship than marriage. As Jesus said, there will be no marriages in Heaven. No paperwork needed, only the commitment one has with the other. Men and women will respect each other and not have any inappropriate earthly desires towards one another. All will be in the family of Jesus Christ. We will keep our identities as people, recognizable as who we were, as evidenced by those who have had NDE experiences who have seen departed loved ones. Every single person will be considered beautiful. No one will be looked at as unappealing or unattractive. We will not age beyond the age of 30. We will be forever young. We will be able to converse with each and every person.

Will we see our beloved departed pets in Heaven? I believe we will. God knows how much we loved them. There will be all kinds of animals in the Heavenly Kingdom, why not our precious little ones also? I can't imagine Heaven without cats and dogs and parrots, etc. entertaining us with their joyful play. They will also be smarter and will remember their masters. We will most certainly be reunited with all the departed humans we have loved. That's a given, and won't that be a grand reunion. I'm not going to delve into what will happen to those we loved who were lost in Christ. I only know the Bible tells us that there will be no more tears and no more sorrow for those who enter the Kingdom of Heaven. This, of course, is a very complicated matter that only God knows how we can deal with this issue, and not for us to speculate. All we can do on Earth is to encourage all our loved ones to come to our Lord Jesus Christ and hope for the best.

Will there be music in Heaven? Yes, the Bible makes references to angels playing trumpets, harps, and also singing. Who makes trumpets and harps you may ask. The answer is, if God can make Earth and the universe he can make a trumpet. All these fascinating questions we all have about musical instruments, buildings, automobiles, highways, electricity, sports, etc. are things that fall under the category of - we'll just have to wait and see. For instance, if we want to speculate about travel, it may be that we will be able to transport ourselves to another location without the use of any mechanical device. There are no clues to any of these questions so we will stay away from hypothesizing.

What we can say is that, in Heaven, we will be much more spiritual than we were on Earth and that is something we will embrace joyfully. As we said earlier, before and during Jesus time, religion played a big part in everyone's life. The focus was on God and then on God and Jesus. Today we have so many activities and distractions we hardly have time to include God, Jesus, and the Holy Spirit in our daily lives. We kind of just fit them into our busy schedules. But in Heaven it will be vastly different for us. We will look forward to our time of worship. As part of our day we will be engaged in service and worship. The connection between service and worship is something we cannot fully understand but will give us great joy and peace. We will worship and sing praises to the Lord. The music will be unlike any we have ever experienced. It will be 'Heavenly'. Picture Jesus teaching to us on a country hillside and the angels singing and playing their harps with trumpets reverberating throughout the land. Worship services will be better than any concert you ever attended. Imagine looking at and hearing Jesus' words of wisdom in person. Wow. The throngs of people will add to the excitement. We will look so forward, with anticipation, to our time with the Lord and with worshiping God. The Bible says we will see God also. That's just mindboggling.

We will be of service to our Lord Jesus Christ as part of our day. Jesus will give us plenty to do and we will have great joy in being of service to him as he also will be of service to us. That's who he is. We also will, most likely, encounter Moses, Abraham, Noah, Paul, and all the Biblical characters, including the twelve disciples.

During the day we will spend time eating and having fellowship with our families and friends. Fellowship will be like having one continuous party, celebrating together forever in Heaven. During the day we will also have rest, real rest, with no thoughts of, 'I need to clean the gutters later,' or 'mow the lawn', or 'get supper ready'. Our minds will not be occupied with thoughts of chores and things that need our attention, making us feel guilty for resting. None of that in Heaven. I'm sure there will also be time for exercise and games, movement of our bodies in some way. Playful activities, no doubt, will be in store for us.

There will be places for privacy, for individual prayers and meditations, and to be with a special group or person. Heaven will provide everyone with all their needs, no one will be left wanting. You're thinking- how can we do all of these activities in a single day? Here are a couple of different thoughts. Our bodies can never get tired, so long sleeping periods are not needed. Periods of rest are sufficient. The Bible says that - after God finished his work, he 'rested' on the seventh day. Now, we will not be like God, but it's just a clue. Another thought would be - is a day just only 24 hours in Heaven like it is on Earth, or is a day in Heaven a week? Is time that important in Heaven? Are clocks necessary? All these thoughts are fun to think about but we don't really know.

Will we have free will in Heaven? This is a very hard question to answer because it's very complicated and not easy to understand. The question is – If we have free will in Heaven then why won't we be tempted to sin? The Bible clearly states that we will sin no more. On Earth free will was given to us by God to allow us to choose good over evil, to do right or to do wrong, to lie or tell the truth, to be honest or deceitful, and so on. We were programmed by God to have these choices for his reasons. Earth is our learning place, our school for the knowledge of good and evil. In Heaven we graduate into a higher learning. In 2 Corinthians 5: 17 (NIV) it says; "Therefore, if anyone is in Christ, the new creation has come. The old has gone, the new is here!" In Heaven, along with the changes in our new bodies, our minds will be programmed in a much different way. As we said earlier, in Heaven we will shed all the human emotions that made us feel bad on Earth, this also includes sinning. In Heaven we will have a superior form of free will that can only choose good. Our thoughts will be pure and righteous. We will no longer be of a sinful mind. We will retain our free will in Heaven but will lose the capacity to sin and do wrong.

Where is Heaven? We have super telescopes and spacecraft's with super long range cameras to look out into space, and beyond, but we have not seen any signs of Heaven. As we said before, Heaven may very well be in another dimension. We have the account of Jesus appearing in a locked room, to the disciples, out of nowhere, in the flesh. The only explanation is that he stepped out of another dimension or he entered by his spirit and then instantly materialized his body upon his entrance. I can't think of any other scenario.

I'm leaning towards another dimension because Heaven is an actual place and not an invisible spiritual place in the sky with our spirits just floating around in the clouds as some believe. It has to be somewhere. Our new bodies have to live on something tangible. Heaven is a real place! So, having said all of this I believe that Heaven is in close proximity to Earth for several reasons. God would want to be close to his children as would the Holy Spirit. The Angels are sent to us as helpers (Guardian Angels) and as messengers. They come, they return. They are a busy bunch. Those who have experienced NDE's have gone there within a short period of time and in an instant their souls returned to their bodies. Some have described it as the snap of your fingers. The fact that Jesus had to come and go to somewhere tells us also he did not travel trillions of miles. Jesus told his disciples "And if I go and prepare a place for you, I will come again, and receive you unto myself; that where I am, there ye may be also." All indicators point to Heaven being close and not in a very far away galaxy. If we can't see it, then Heaven has to be in another dimension. I don't think that is stretching our minds, considering what we already believe in and know about the wonders of God. Why not this? People have been talking about other dimensions for ages. Where did the idea of other dimensions come from and why? It's another mystery of many. There is nothing our minds can perceive that you can say is impossible for God. Earth is a fascinating place for us to live, but what is really wonderful for the believer is to know that there is an even greater place than Earth that awaits us, and not fearing the passing through into a glorious Heavenly Kingdom where our lives will really begin.

Chapter Seven

FINAL THOUGHTS

Some further thoughts on evolution. The fact that the most intellectual of all pre-human species (Neanderthal and Cro-Magnon man) disappeared from the planet during the Ice Age makes it unlikely that any hominins survived the Ice Age. None of these hominin types ever became human as scientists would like us to believe. Also, the fact that there is no direct evidence to support their theories, like the missing link fossil that connects hominins to modern humans, convinces us that Adam and Eve were the first 'real' humans on Earth, created by God. Adam and Eve were unique in their makeup and far more intelligent than the hominins that roamed the planet for millions of years. Scientists want to call anything that walked on two legs early humans. Early humans were Adam and Eve. Even Cro-Magnon and Neanderthal had no lineage to humans. According to the Bible there wasn't any life left on the planet anyway around 4500 years ago after the flood of Noah, except Noah, his family, and the animals that were on the arc.

After the flood of Noah, sudden and real technology 'exploded' in the world in just a few thousand years to what we enjoy today. This could only have come about by highly intelligent beings, and not by the excruciatingly slow development of hominin 'tool making' over millions and millions of years. We can't deny that there were all kinds of creatures on this planet, including the dinosaurs. God is responsible for any 'living creature' that ever roamed the planet. He, and only He, created everything. Scientists can try to prove otherwise by digging and digging, but it will be to no avail. Creationists have all the proof they need, it's in the Bible!

The mere mention of God must drive evolutionists up the wall, but what they don't realize is that without him they wouldn't exist and be able to do the work they do to deny the very one who has created them. Irony!

The fact that some ancient (primitive) cultures still exist in the world today does not mean that they evolved here from early times. I mention this only because it's a topic I have encountered in the past from those who try to explain the vast differences between human cultures. I apologize if this subject does not sound politically correct but it's out there never the less. The answer is that God has created human diversity as he did with all of his creations. Not all animals and plant life are the same and neither are humans. The order of animals and humans in the world cannot be questioned and is far beyond our understanding.

Final thought regarding the 'Big Bang' and evolution. Neither theory or hypothesis brought us to where we are today. It could only have happened through 'intelligent design' by God the Creator.

Now I'd like to talk about the 'unbeliever'. In this entire book I have not referred to them as 'atheists'. To me that's an offensive word of sorts. It's a negative connotation. Each person has the right to believe in what they choose. I had a good friend that was an unbeliever. He was a great person. He was honest, kind, generous, and fun to be with. He never said a bad word about anybody. He was more Christian-like than many Christians I know. I could never figure out how someone could be so morally good without having God in his life. I always thought unbelievers could do anything they want without fearing reprisals, other than penalties for breaking the law. I avoided any religious conversations with him out of respect for his beliefs and just enjoyed his company. He has since passed and I often think about him with sadness, wondering what his ultimate fate turned out to be. Did this wonderful man wind up in hell or did the Lord just let him simply pass, ending his existence? Does the Lord make these finite distinctions between the very evil and the good who are just not believers? This is a very perplexing question that there is no answer to.

    What's the psychology behind the unbeliever's thought process? Is it bitterness, something terrible that happened to them or their loved ones, or is it simply that they cannot conceive of anything other than only what we can accomplish ourselves here on Earth? Is it that their minds cannot perceive of a higher and greater power? Is it just too good to be true for them, or did they just simply inherit their belief system from their parents? We all know enough of what the Bible says. It all comes down to believing what the Bible says, or not believing what the Bible says. But it's more than that too. Before I really started reading the Scriptures as a young boy I had a profound belief in God, Jesus, and Heaven, like it was ingrained in me as a toddler. I realize now, of course, that the Holy Spirit was in me from the time I was baptized as an infant. Also, having Catechism and confirmation gave me the proper training to move in the right direction. I thank my parents for being diligent enough to push me through the completion of this entire process. I think parents today need to think about this and take their children through the religious processes required by their respective religious institutions as the Jewish do with Bar and Bat Mitzvahs for their children. It's important for parents to do their part while the children are young. The rest will be up to them as they march into their future.

I believe in baptizing babies, if only for the reason that we can't leave it up to our children to go get baptized as adults themselves. We know that most will never get around to it, and that's a sad fact. Even if you force them into baptism when you think they're old enough to understand does not guarantee that they are or will be believers, so why not just get them baptized as infants. I would love to take a poll of unbelievers to see how many were never baptized. I wonder if there is any correlation there. Many more people believe in God and Heaven than those who do not believe. It has nothing to do with intelligence. People who believe just simply seem to know that there is a God, even those who never read one word in the Bible. What is it? It's the work of the Holy Spirit. That's why baptism is so important.

Some thoughts for the 'believer'. Many believers do not attend church. They convince themselves that believing and saying prayers at home, with some also reading the Bible, is enough so that they really don't need to go to church. But that is 'not' what God wants. Regarding worship, the Bible is very clear that the Lord wants us to gather together as a family to worship and praise him and to hear the Scriptures and the word of God, especially for the children's sake. The Christian Church, or any church for that matter, cannot grow without active participation by its believers in attending worship services. It can't be done individually at home. The Lord's church is only as strong as the people in it, not only to worship, but to be active and involved within the church community, to do good and to spread the gospel.

In Matthew 18:20 (KJV) - Jesus says: "For where two or three are gathered together in my name, there I am in the midst of them." What Jesus is saying is that he wants people to come together as a group (not just as one) in his name and he will be present. Here it is clear that Jesus is encouraging us to 'gather together' to worship, but at the same time he is not discouraging us from also praying in private. He requires us to do both. John 4:23 (KJV) – "But the hour cometh, and now is, when the true worshippers shall worship the Father in spirit and in truth: for the Father seeketh such to worship him."

Believers, give serious thought to supporting your local churches as we all know church attendance has been declining, especially in certain parts of the country. Many churches are closing their doors for lack of support. One excuse that some give for not attending church is that 'the churches have become corrupted'. Neither the churches nor the religious entities are corrupted. It's people who corrupt. Every organization or entity has people in it who are dishonest, whether it's schools, financial institutions, politics, doctors, lawyers, police, charities, churches, and so on, and that will never change. Take the police for instance. There are, undoubtedly, over a million law enforcement personnel in this country, and because of a small percentage of corrupt police officers, people want to indict the whole system as 'the police are bad'. The police are good and they stand between us and the criminal element in our society, putting their lives on the line for us every day. Try having a society without a police force. All we can do in any organization is to identify those who are bad and remove them from the organization. Same goes for the churches.

As believers, sometimes we are misguided in how we try to spread the gospel to others. We say things like 'you need to have God and Jesus in your life', or 'God will punish you if you continue to do the things you do', or 'you need saving', or you need to do this or do that. These kinds of comments put yourself on a higher plane, feeling the need to advise others on how to live. The term used here is, 'holier than thou'. People do not like to be given advice on how they should live their lives or to be judged. Jesus said "Do not judge, or you will be judged. For in the same way you judge others, and with the measure you use, it will be measured to you."

A better way to bring attention to the Lord is how some athletes do it when they are praised by reporters for playing well on the field or on the basketball court. They'll say things like "All the glory goes to God" or "God was with me." I heard LeBron James, arguably the greatest sports player in the world today, credit the "Man above" after the Cavaliers won their first championship. There are many other examples of comments like this from other sports stars as well. Subtle comments about God and Jesus in our everyday lives are far more effective than preaching to people. Let's leave that up to our ministers, etc. on Sunday mornings. I know believers have good intentions in spreading the gospel to others, but in today's world we need to take a different approach. For example, if you just had surgery, you can say 'I thank God for pulling me through' or if you had an accident and didn't get hurt, you can say 'God was watching out for me'. Even when taking a test, you can say 'I prayed to God last night that I would do well and I did'. You can do this for almost any situation you have, but you have to be smart about it. Don't overdo it. Personal testimonies, like the examples I've given, giving God and Jesus credit for your successes, or mentioning that you prayed for something, is a great way to keep the Lord's name in the public light. You can't advise people about the Lord and the need to go to church to a group of people in a social setting, otherwise they are going to think to themselves 'who wants to hear this now'. Of course there are times when a close loved one needs guidance, that talking about the Lord in a loving way can be very helpful.

Back to the unbeliever. I don't know how many people, who do not believe, will read this book. The agnostic (doubter) may. It would be great for me to be able to, at least, provoke some thought about the existence of the Creator to those who doubt or don't believe. I know that the unbeliever thinks - 'how can Christians actually believe all this Bible Stuff? But there are billions of us. Can we all be wrong? I, myself, can't believe that the unbeliever doesn't believe, with all the evidence that's available to them. I want you, as the doubter and unbeliever, to consider the prospect of missing out on the great opportunity that Jesus' life has afforded us - to live beyond our earthly existence in a far greater place than Earth itself, forever! It's a 'big' price to pay for your unbelief or doubts. Now, I know it's not easy to make your mind believe in something that you don't believe in. It will take a lot of hard work. But if you try, you will really please the Lord and he will help you.

Jesus' parable of the lost sheep - Luke 15: 4-7 (NKJV) - "what man of you, having a hundred sheep, if he loses one of them, does not leave the ninety-nine in the wilderness, and go after the one which is lost until he finds it? And when he has found it, he lays it on his shoulders, rejoicing. And when he comes home, he calls together his friends and neighbors, saying to them, "Rejoice with me, for I have found my sheep which was lost!" I say to you that likewise there will be more joy in Heaven over one sinner who repents then over ninety-nine just persons who need no repentance." I know that there are a lot of you very nice people out there. You're welcome to join us. We'd love to have you! If writing this book can bring just one who is lost back to the Lord, then that will be my reward.

www.ingramcontent.com/pod-product-compliance
Ingram Content Group UK Ltd.
Pitfield, Milton Keynes, MK11 3LW, UK
UKHW022216230426
12048UKWH00016BA/875